PLANETS FOR MAN

STEPHEN H. DOLE

ISAAC ASIMOV

Planets for Man was originally published by Random House in 1964. This RAND edition reflects the original layout.

Library of Congress Cataloging-in-Publication Data

978-0-8330-4226-2

The RAND Corporation is a nonprofit research organization providing objective analysis and effective solutions that address the challenges facing the public and private sectors around the world. RAND's publications do not necessarily reflect the opinions of its research clients and sponsors.

RAND® is a registered trademark.

Cover Design by Peter Soriano

Published 2007 by the RAND Corporation
1776 Main Street, P.O. Box 2138, Santa Monica, CA 90407-2138
1200 South Hayes Street, Arlington, VA 22202-5050
4570 Fifth Avenue, Suite 600, Pittsburgh, PA 15213-2665
RAND URL: http://www.rand.org/
To order RAND documents or to obtain additional information, contact
Distribution Services: Telephone: (310) 451-7002;
Fax: (310) 451-6915; Email: order@rand.org

Preface

The aim of this book is ambitious in that it tries to suggest answers to some very basic questions about the ultimate goals of space travel. Such projections are risky, considering that our knowledge of the universe is still so incomplete, but, for the adventuresome of mind and imagination, who will, we hope, accept it in the spirit in which it is offered, it will open glimpses of possible new horizons for future generations of human beings.

Many of the subjects discussed herein are highly controversial for there is no general agreement among scientists on numerous questions. As an example, on concepts of the formation of stars and planets (cosmogony), there are at least as many theories as there are writers on the subject. No two of these theories agree in all their basic details, and even where there is some agreement now, it is possible that drastic revisions will have to be made in the future as more evidence is accumulated that requires the abandonment of some ideas and the introduction of new ones. This has happened many times in the past and undoubtedly will continue to occur until all the pertinent facts have been gathered and all the pieces fall together in a coherent and entirely convincing manner.

No attempt has been made to present all points of view on controversial questions or all the current theories that differ from those advanced here. To have made the attempt would have resulted in a hopelessly long and complicated book. Those desiring to see more detailed technical substantiation of the ideas presented might wish to consult the book by one of the authors, *Habitable Planets for Man*, on which the present one is based. The numerous graphs, tables, and specific references to published papers and other bibliographic material contained in *Habitable Planets for Man* have been omitted from the present book, which is meant for the general reader. For those who wish further information, a list of Related Reading Material containing some of the more general works consulted has been included at the end of the book.

S. H. D. and I. A.

Contents

PLANETS FOR MAN

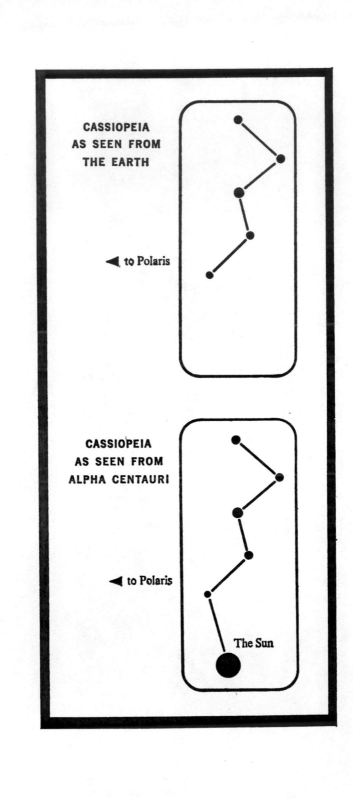

CASSIOPEIA
AS SEEN FROM
THE EARTH

◀ to Polaris

CASSIOPEIA
AS SEEN FROM
ALPHA CENTAURI

◀ to Polaris

The Sun

The Coming
Search

The Brightest Star in Cassiopeia

The star Alpha Centauri is our nearest neighbor in space beyond the Solar system.

By Earthly standards, "nearest neighbor" is not very close at all, for Alpha Centauri is 25 trillion miles away. Light, which travels 186,000 miles per second, takes 4.3 years to travel from Alpha Centauri to ourselves. We can therefore say that Alpha Centauri is 4.3 light years away. Yet such is the vastness of the Universe that no star is closer than Alpha Centauri, while billions of stars are much, much more distant.

Let us look, then, more closely at this far-distant nearest neighbor of ours. Because of its nearness, Alpha Centauri, though not one of the very luminous stars, is the third brightest star in the sky. (It is not visible, however, from most of the North Temperate Zone, so Americans and Europeans do not ordinarily see it.)

Alpha Centauri is actually a triple star system. The two larger components of the system, Alpha Centauri A and Alpha Centauri B, revolve around each other in a period of 80 years, at a mean separation of about 23 astronomical units. (An "astronomical unit," or "A.U.," is the average distance from the Earth to the Sun—about 93 million miles —so that 23 A.U. represent about 2.1 billion miles.) The third member, Alpha Centauri C, some 10,000 A.U. (about 1 trillion miles) removed, revolves around the pair A-B in a period of the order of a million years.

The two stars Alpha Centauri A and Alpha Centauri B are each very much like our own Sun, and either or both might conceivably possess a habitable planet like our Earth. As of now we have no way of actually telling whether such a habitable planet circles Alpha Centauri A or Alpha Centauri B, but suppose one does. We might then visualize an intelligent creature on such a planet surveying his night sky as we survey ours.

The intelligent Centaurian, studying the stars, would see, almost unchanged, the familiar constellations we see from the Earth. His star, you see, is so close to ours as compared with the vast scale of the Universe that the general view of the night sky is just about the same there as it is here. He would see the Big Dipper, Orion, and the constellations of the Zodiac.

But there would be one outstanding addition. Instead of the five bright stars making up the familiar zigzag of the constellation Cassiopeia, as we see it, our Centaurian would see six. The sixth star would extend the zigzag pattern one more step and it would be six times brighter than any of the other five. That sixth star would be our Sun.

As far as our Centaurian would be able to see, there would be nothing particularly remarkable about this star except that it would be brighter than most. Barring that, it would look very much the same as many others within his view. If his science were no more advanced than ours, he would have no way of detecting the fact that what was to him the brightest star in Cassiopeia possessed a life-bearing planet in orbit around it. He could only speculate and wonder, as we speculate and wonder about Alpha Centauri—and about other stars.

The Promise of Space

And why bother with all this speculation and wonder? Until very recent years, the only excuse that could possibly have been offered was that men were irremediably curious and had to wonder for the sake of wonder itself. But now we have a

more practical reason to offer, for mankind stands
at the threshold of space flight.

To be sure, the plans now being made for the
manned exploration of the space near the Earth
may be likened to the first efforts of a baby who
has not yet learned to creep. It is a learning process
with only short-term goals as yet. Our first falter-
ing steps will take us to the Moon. More confident
steps will lead us on to explore the surface of Mars
and the cloud cover of Venus. Later on, we will
reach the other planets of the Solar system.

None of these worlds would be comfortable
for us, but we need not be too distressed over this
fact for our exploration of our sister planets will
represent but the barest beginning. Consider
where the process will lead us when we have
really learned how to traverse space with assurance
and ease!

This book takes the long-term viewpoint. It
looks forward to a time when man will be able to
seek and find habitable planets beyond the Solar
system. To be sure, we cannot do this now and we
cannot even see how the vast spaces between the
stars can be bridged at all, within the scope of
our present limited knowledge. However, we are
barely started on our upward climb as self-aware
and knowledge-seeking creatures. There is much
more to learn than has already been learned, and
in view of the rapid and accelerating strides of
science and technology over the past 50 or 60
years, it does not seem outlandishly optimistic to

suppose that practical methods for leaping the
interstellar gulfs will be developed in the not-too-
distant future.

Therefore, let us assume that this is so and
consider not the means of reaching the stars, but
our destinations. Once we set out on our journey
to the stars (by whatever means), let us ask the
question: Toward which stars ought we to travel?

Does it make a difference? Certainly it does,
if we but consider what the goals of manned space
flight might be.

One goal, most certainly, would be to increase
our knowledge and understanding of the Universe:
to attempt to learn about the origin of matter, of
galaxies, even of life. A second might be to find
new sources of energy and raw materials. But
surely there is a third purpose that ought to prove
the most attractive goal of all: that of finding new
habitable planets.

As it is now, one planet, the Earth, supports
the entire human race. One planetary catastrophe
could completely destroy us. But if the human
race were living on a number of planets scattered
around the Galaxy, its immortality would be assured. What's more, the opportunities for variations in culture and in outlooks would be vastly
multiplied, and the interactions among the multi-
tude of human subgroupings might vastly accelerate the over-all progress of mankind toward ultimate mastery of the physical universe.

But even if the possibility of interstellar flight

is assured, it will remain an expensive and difficult proposition for a long time. We will not be able to afford to wander through interstellar space at random. We would have to guide our ships with care and thought in this great search that is to come. It is not too soon, even now, to begin to ask ourselves: Where ought we to begin in the coming search for habitable planets? And what is the probability of finding them among the nearby stars?

This book tries to give some reasonable answers to these questions.

Approach to the Problem

The search for other habitable planets cannot be direct. Objects of planetary size at the distances of even the very closest stars cannot be detected optically under even the most favorable conditions. The use of the Palomar Mountain 200-inch telescope, the world's largest, could not help us. Lacking direct evidence, then, we will have to use indirect reasoning.

The task will be all the harder since we will set unusually stringent conditions. Many astronomers have speculated before now on the subject of life on other planets but only in a relatively vague and qualitative manner. We, however, will aim at obtaining as exact a set of criteria for habitability as is possible. We will try to establish reasonable quantitative limits for environments favorable to man and then calculate the prevalence of habitable planets that meet those limits.

To do this, we will have to have a starting point, and ours will be the assumption that our Solar system is not an extraordinarily rare assemblage of bodies, but that it is rather a typical planetary system, and that its members can be treated as a good, although not numerically large, sample of the sorts of bodies that exist in the proximity of other stars. It is only fair, then, to begin by asking if a habitable planet, other than the Earth itself, exists in the Solar system.

To that the answer must be: No!

The reasons for this negative answer will be considered later in detail, but here are the conclusions in brief:

Mercury is too small to retain a breathable atmosphere. It is so close to the Sun that the Sun has apparently stopped its rotation. Its surface no longer has an alternation of day and night; its sunlit side is therefore too hot, and its dark side too cold.

Venus, although of the right size, is close enough to the Sun so that its rotation, too, has apparently been slowed drastically. Its atmosphere probably contains too much carbon dioxide to be compatible with human life, and its surface temperatures are almost certainly too high.

The *Moon*, our satellite, is at the same distance from the Sun that the Earth is. It is therefore at the correct distance from the Sun for habitability, since the Earth itself quite obviously is. Nevertheless, the Moon is too small to retain an atmosphere or even to produce one suitable for man. Because its rotation with respect to the Sun is slow, its days are too hot, and its nights are too cold.

Mars is also too small to produce or retain an atmosphere suitable for human beings. It is far enough from the Sun so that even if it were massive enough, its average surface temperatures would probably still be too low for it to be considered habitable.

Jupiter, Saturn, Uranus, and *Neptune* are so massive that they have retained enormously thick atmospheres, consisting mainly of hydrogen and helium, which are unbreathable by man. The satellites of the major planets are too small and too cold; so are the asteroids. *Pluto* is also too cold.

Perhaps it may seem to you that we are too cavalier in this rapid dismissal of the bodies of the Solar system as non-habitable. Surely, Mars at least is often spoken of as a possible abode of life, and men do plan to establish colonies on the Moon. Let us, then, explain precisely what we will mean by the word "habitable" when used in this book, and point out, in particular, that as used here, it is not synonymous with "life-bearing."

Planets might be habitable in a variety of different senses. Planets could, for instance, support some unknown form of life with a chemistry

basically different from that of life on Earth. We
might, for example, speculate about organisms living in seas of liquid ammonia on Jupiter or breathing gaseous sulfur on Mercury.

Again, planets might be inhabited by carbon-
based "life as we know it," but under conditions
that could support only microscopic forms of such
life, or only specialized forms specifically adapted
to extreme environments. Mars might belong to
this group of planets. There is indeed the possibility that a form of plant life exists on the planet,
but if so it must, to all appearances, consist only
of very small, or even microscopic, organisms, because of the extreme scarcity of water on the
Martian surface.

Then, too, planets might be made habitable
for man himself at the price of extensive feats of
engineering. The atmosphere or surface of a
planet might be remodeled so that people could
live there in large numbers. On Mars, or even on
the Moon, water might be obtained from surface
rocks, food might be grown under glass, and
people could live in hermetically sealed "hot-
houses" containing breathable atmospheres. Such
highly artificial conditions, involving underground
cities or domed cities on the surface, would probably never be entirely independent of supplies
from Earth and would be vulnerable to even temporary breakdowns in the complex technology
serving to support the system.

No planets of any of these types are to be

considered "habitable" for purposes of this book. We will define a habitable planet as one on which large numbers of people can live comfortably and enjoyably, without needing unreasonable protection from the natural environment and without dependence on materials brought in from other planets. In short, the vast majority of habitable planets, by our view, will turn out to be worlds very much like the Earth.

It is in this sense that none of the bodies of the Solar system, other than the Earth itself, is habitable. If other habitable planets exist, they can be found only in the neighborhood of stars other than our Sun.

Our plan of attack on the problem we have set ourselves, then, falls into three parts:

1. We will first describe the environmental conditions required to make a planet habitable.

2. We will then work out the combination of astronomical circumstances that will produce these conditions.

3. We will finally estimate the probabilities that the necessary combination of astronomical circumstances will be found elsewhere in the Galaxy and, in particular, where they might be found among those stars relatively close to the Sun.

Let's begin, then, with the first part.

2

Human
Requirements

Temperature

In itemizing the specific requirements necessary for planets to be habitable for man, the matter of temperature is a reasonable starting point. While it is true that human beings can endure brief periods of extreme heat and cold by using various kinds of protective clothing and other insulation, it is also true that there is a range of temperature that human beings prefer for their everyday existence and that variations outside the zone of maximum comfort are of constant interest to us. (We all know what the chief topic of conversation is bound to be on days that are unusually hot or unusually cold.)

Virtually all the world's human population

lives in regions where the mean annual temperature is between 32° F. (0° C.)* and 86° F. (30° C.). Fully 18 per cent of the Earth's land area (including Antarctica, Greenland, the Arctic islands, etc.) has a mean annual temperature of less than 32° F. (0° C.) and is, by our definition, as uninhabitable as the Moon, though easier to reach. In fact, 95 per cent of the human population is crowded into the 65 per cent of the Earth's land area that has a mean annual temperature between 40° F. (4.5° C.) and 80° F. (27° C.).

It is, of course, not only the human desire for comfort that dictates such a narrow temperature range; it is also the fact that these temperatures are best tolerated by the agricultural crops and domesticated animals on which man depends for food. To be sure, some species of plants and animals can withstand more or less prolonged exposures to very high or very low temperatures, and a few species have become adapted to hot or cold environments, but most of the plants and animals important to man as sources of food and as suppliers of oxygen require temperatures between 32° F. (0° C.) and 86° F. (30° C.) for survival and active growth.

In addition to the limitation that involves the

---

* The Fahrenheit scale (°F.) is in common use in the United States and Great Britain. Most of the rest of the world, and scientists everywhere, including American and British scientists, use the Celsius (°C.), also called the Centigrade, scale. We will present all temperatures in this book in both scales.

mean annual temperature, there is also a limitation introduced by the daily temperature extremes experienced at the warmest and coldest seasons of the year. Thus, an area with an annual mean temperature of 60° F. (15.5° C.) may experience a daily mean temperature of 80° F. (27° C.) in midsummer and 40° F. (4.5° C.) in midwinter. Such a temperature range is clearly within habitable limits. It is conceivable, though, that another area may have a daily mean temperature of 140° F. (60° C.) in midsummer and −20° F. (−29° C.) in midwinter. The annual mean temperature might still be 60° F. and yet the area would be uninhabitable. The average, in other words, is never safe, unless we know that the balancing extremes that contribute to the average are not *too* extreme.

On the basis of human tolerances, then, and assuming that human beings would not be comfortable having to stay indoors constantly over long periods of time in the more extreme seasons, we can fairly decide that mean daily temperatures of 104° F. (40° C.) and 14° F. (−10° C.) at the hottest and coldest seasons would represent reasonable limitations. If this seems too conservative, remember that the figures are averages over the day. In the hot season, mid-afternoon temperatures might, even with our limitation, rise to a maximum of 120° F. (49° C.) and in the cold season, the temperatures at dawn might be as low as −5° F.

(—20.5° C.). Exposure to temperatures of 120° F. (49° C.), at even moderate humidities, would result in the danger of induced fever ("hyperthermia") in the space of an hour or so, while exposure to temperatures of —5° F. (—20.5° C.), even when people are warmly dressed, involves danger of frost-bitten extremities after a couple of hours.

On any given planet, the temperature range would vary from region to region, making for habitability in some places and not in others. Even the Earth itself, as we said above, has its uninhabitable regions, so that only about 80 per cent of its land area is suitable for us. It is not likely that many planets can do much better than this, but they could easily do much worse. Naturally, if only isolated fragments of a planet's surface are habitable, it is scarcely worth man's while to try to colonize it. For us to consider a planet as habitable it would seem right, then, to expect at least a reasonable fraction of its surface area (say, 10 per cent) to be habitable.

To summarize the temperature requirements, then, we specify that a planet is habitable only if the mean annual temperature of at least 10 per cent of its surface lies between 32° F. (0° C.) and 86° F. (30° C.), and if the highest mean daily temperature during the warmest season is not higher than 104° F. (40° C.), while the lowest mean daily temperature of the coldest season is not lower than 14° F. (—10° C.).

Light

That portion of the electromagnetic spectrum visible to the human eye, which we call light, is a wave-form in which the length of the individual waves ("wavelength") varies from 380 to 760 millimicrons. (A millimicron, abbreviated mμ, is a billionth of a meter or a 25-millionth of an inch.)

By using very intense artificial sources, one can stretch the limits of human vision somewhat more widely, from 310 to 1050 mμ. Lying generally within this slightly wider range of wavelengths but mainly enclosed within the narrower range, we find also the vision of other animals, the oriented movements of simple animals toward light or away from it, the bending of plants toward light, and so on. Most important, it is light in this region that stimulates photosynthesis, the chemical reactions whereby green plants convert carbon dioxide and water into carbohydrates and free oxygen, supplying the food and replenishing the atmosphere for the entire animal kingdom. It is quite likely that these same limits are applicable throughout the Universe, wherever man could find a comfortable place to live.

We must also concern ourselves with the intensity of illumination ("illuminance") required for human vision and for other phenomena vital to human life. Illuminance is commonly measured in lumens per square centimeter, abbreviated "lu/cm²." Without trying to define the lumen, we can set a standard by pointing out that the maximum illuminance due to direct and scattered sunlight at the surface of the Earth is about 15 lu/cm².

The range of illuminance over which human vision can operate is wide indeed. Human beings can see well enough to walk around with reasonable assurance if the illuminance is as low as 10^{-9} (one-billionth) lu/cm². The absolute lower limit of naked-eye detection of a ray of light shining directly into the eye out of very dark surroundings is lower still, about 10^{-13} (one 10-trillionth) lu/cm². This corresponds to a star with an apparent visual magnitude of +8,* although under the best typical viewing conditions, it is difficult to see stars fainter than magnitude 6.5.

There is an upper limit of illuminance too, of course. A man looking directly at a point source of light would find matters intolerable if the illuminance were of the order of 0.05 lu/cm² or

---

* By a convention as old as the Greeks, the apparent brightness of objects in the sky is measured by a series of numbers. The "apparent visual magnitude" of the dimmest stars ordinarily visible to the naked eye is set at +6. Brighter stars are +5, still brighter ones +4, and so on. The brightest stars in the sky have magnitudes ranging about +1 ("first-magnitude stars"), though the very brightest go on to magnitudes of 0 and even of −1. The full Moon, on this scale, has a magnitude of −12, and the Sun one of −27. Each successive step on the magnitude scale represents a 2.5 multiplication of brightness. Thus, a star of magnitude +2 is 2.5 times as bright as one of magnitude +3 and 2.5×2.5 or 6.25 times as bright as one of magnitude +4, and so on.

more. This corresponds to an object with an apparent visual magnitude of about −21. Thus, the full Moon, with a magnitude of −12, is only 1/4000 as bright as this maximum and can be looked at with impunity. The Sun, with a magnitude of −27, is 250 times brighter than this maximum and cannot be looked at directly without damage to the eye.

The upper limit can be raised if one deals with over-all illuminance rather than with light entering the eye directly. Here, many complicating factors must be considered, such as the reflecting power of surfaces and objects in the vicinity, the presence of shade and shadows, and so on. Even ordinary levels of illuminance due to sunlight at the Earth's surface become intolerably high when one is surrounded by material, such as fresh snow, that reflects most of the light diffusely. It is this that gives rise to the well-known phenomenon of snow-blindness. Under the best and most moderating of conditions, however, an over-all illuminance of 50 lu/cm^2 (three times the illumination level of bright sunlight) may be taken as an upper limit.

The limits, set by the eye, of 10^{-13} to 50 lu/cm^2 are extraordinarily broad, but other factors cut down the permissible range drastically. Daily illumination intensities for active growth in green plants, for instance, must fall between certain more limited extremes. If the illuminance is too low, active photosynthesis cannot proceed at a rate high enough to be useful; and if it is too high, growth is slowed down by what has been termed "solarization." The limits have not been clearly established, but they may be set at approximately 0.02 and 30 lu/cm^2. The highest growth rates for terrestrial plants are encountered at intermediate levels of illuminance. For some common species of algae, for example, the highest growth rates were found in the approximate range of 0.3 to 3.0 lu/cm^2.

Thus, illumination requirements are set primarily by the needs of plants and not by the needs of vision. For habitability, the range of illuminance, during the daylight hours of the growing season, must lie between 0.02 and 30 lu/cm^2.

Another factor of great importance to the growth of plants is the periodicity of illuminance; the length, in other words, of day and night. Especially in the temperate regions of the Earth, plant growth cycles are determined by the relative or absolute lengths of days and nights, as well as by temperature patterns.

Fortunately, however, the matter of illuminance does not complicate the habitability picture as much as one might suppose. On a planet such as the Earth, the source of both light and heat is the Sun. In general, any effect that would alter the quantity of heat emitted by the Sun would also alter the quantity of light emitted by it. It follows, then, that if a planet has a habitable mean annual temperature, as the Earth does, it also, in all probability, experiences Earth-like levels of illuminance.

Furthermore, we shall see that if a period of rotation is too long, daily temperature extremes make the planet uninhabitable. The days grow too hot, the nights too cold. If the rotation is such that the planet meets the temperature requirements, then the periodicity of illumination will also meet the requirements for habitability.

On the whole, then, we can conclude that a planet that is habitable in terms of temperature is also very likely to be habitable in terms of illuminance as well.

Gravity

The gravitational intensity on the surface of the Earth imposes a pull of 1 g ("g" is an abbreviation of "gravity") on all objects, including human beings, upon that surface. Clearly, men can and do endure this for indefinite periods.

Scientists cannot vary the gravitational field directly in the laboratory, but they can produce forces through acceleration (by means of a rapidly turning centrifuge, for example) that resemble high gravitational fields to a considerable degree. Such centrifugal forces, involving a large angular velocity, are not strictly comparable in all respects to those produced by a massive planet with low angular velocity, but they are as close as we can get.

Experiments with human beings in large centrifuges have shown that relatively high levels of acceleration can be tolerated by some people for brief periods of time without permanent damage. For example, accelerations of the order of 5 g (producing the effect of making each part of the body seem to weigh five times what it ordinarily does, without any corresponding increase in muscular strength) can be tolerated by a seated man, not protected by special gear, for about 2 minutes. Exposed for longer than that, he experiences a "blackout," that is, a loss of vision caused by an inadequate supply of blood at the eye level, since blood cannot be pumped upward by the heart in sufficient quantity, over longer periods, against its own five-times-normal weight.

An acceleration of 4 g can be tolerated for about 8 minutes, while 3 g have been tolerated for as long as an hour by some subjects in several experimental runs. The subjects were seated and immobilized, however; they were not walking around or otherwise functioning in an everyday manner. At the conclusion of the 3-g experiments, the subjects reported quite pronounced muscular fatigue.

Other experiments, conducted in 1947 at the Mayo Clinic, give a sharper idea of the limitations imposed by increased gravitational fields. In these experiments, five human subjects were timed to see how rapidly they could scramble, creep, or crawl across the end of the centrifuge gondola, a

distance of 7.5 feet, under various imposed accelerations. Ordinarily, the time required (averaged
over five subjects) was 1.5 seconds. At 1.41 g,
the time required was 4.9 seconds; at 2.24 g it was
9.4 seconds; and at 3.16 g it was 15.8 seconds.
The subjects were also timed to see how quickly
they could put on a standard parachute at various
g levels. The average times required by three subjects were 17 seconds under ordinary conditions,
21 seconds at 1.41 g, and 41 seconds at 2.24 g.

It seems fair to conclude that the work required to perform various acts becomes excessive
above approximately 2 g, and that life would become burdensome over extended periods at such
a level of gravity. One might conclude, in fact,
that few people would choose to live indefinitely
on a planet where the surface gravity was greater
than 1.25 or 1.50 g. It is true that many people
who are 25 to 50 per cent overweight (and who
therefore experience, after a fashion, the equivalent of 1.25 or 1.50 g) live normal lives and manage to accomplish as much as, or more than, many
people whose weights conform more closely to
the standards for their heights and ages. On the
other hand, it is also generally true that physical
activity is more exhausting to people who are
carrying an excessive burden of fat, and it is better, on the whole, from the standpoint of both
health and performance, that they not do so.

Animal experiments have pointed to similar
conclusions. Both plants and insects apparently

can tolerate extremely high g levels, even up to thousands of g's, but this is certainly not so for vertebrates. Chickens grown in centrifuges for extended periods of time were able to survive prolonged exposure to accelerations up to 4 g, but lost weight unless the acceleration was less than 2.5 g. In these fields of high g, the heart rate increased and the rate of respiration decreased. The life span of small animals also appears to decrease at gravitational forces higher than 2 g (although some mice showed increased life spans at levels between 1.5 and 2.0 g).

There does not seem to be a corresponding lower gravitational limit on the tolerances of human beings; that is, there is no conclusive evidence that a certain minimum level of gravity is required for their normal physiological functioning. Manned orbital flights have shown that human beings will tolerate o g (complete weightlessness) for at least 5 days. To be sure, a planet with a low gravitational field may be non-habitable because it is incapable of retaining an atmosphere but not necessarily because of any direct harm wrought on human beings by such a field.

For habitability, then, we need only specify that a planet have a gravitational field of less than 1.50 g.

Atmospheric Composition and Pressure

A habitable planet must, of course, have a breathable atmosphere, and such an atmosphere can be rather completely specified in terms of its component gases and their concentrations, or partial pressures.

As far as we know, the only essential ingredients of a breathable atmosphere are oxygen and minor amounts of water vapor (though a couple of other components are necessary for purposes other than human respiration). The requirement for oxygen may be judged from actual conditions on the Earth. The pressure of Earth's atmosphere at sea level (at or near which most human beings live) is equal to that of a column of mercury (chemical symbol, Hg) 760 millimeters, or 30 inches, high. Normal air pressure, therefore, is said to be 760 mm. Hg; this quantity can also be called "1 atmosphere."

Oxygen makes up 0.209 of the volume of the atmosphere; that is, just about a fifth. The partial pressure of oxygen is therefore 0.209 atmosphere, or 159 mm. Hg. It is necessary, however, to make a correction in this figure because as air is inhaled, it is also humidified in the nasal passages and throat, so that by the time it reaches the interior of the lungs, it is normally saturated with water vapor at body temperature. This takes up some of the air volume and replaces a bit of the oxygen that would otherwise be present. The result is that the *inspired* partial pressure of oxygen (that entering the lungs) is lower than that in the atmosphere itself. Under normal conditions, the inspired partial pressure of oxygen is 149 mm. Hg. At this inspired partial pressure men can and do live comfortably through extended lifetimes. This can also be done at lower inspired partial pressures.

The lower limit of inspired oxygen partial pressure is approached by the inhabitants of a mining settlement at Aucanquilcha in the Chilean Andes, situated at an altitude of 17,500 feet above sea level. This is said to be the greatest altitude at which men are known to live permanently. It is considerably higher than the environment of the Tibetans, most of whom reside and work their land at altitudes between 12,000 and 16,000 feet.

At 17,500 feet, the inspired partial pressure of oxygen is about 72 mm. Hg, yet the miners of Aucanquilcha lead very strenuous lives and appear to be completely acclimated to the low level of oxygen pressure. To reach the entrance of the mines where they work, they climb 1500 feet each day to an altitude of 19,000 feet, where the inspired partial pressure of oxygen is 68 mm. Hg. And even these conditions may not represent the ultimate lower level of oxygen pressure that can be tolerated indefinitely by some men. The suggestion has been made by some mountain climbers that life can be carried on normally for indefinite periods at 23,500 feet, a height at which the inspired partial pressure is a mere 53 mm. Hg, a little more than one-third the value at sea level. We will split the difference, however, and set the lower limit of inspired oxygen partial pressure at 60 mm. Hg.

The upper limit of inspired oxygen partial pressure has been found, experimentally, to be approximately 400 mm. Hg. This is equivalent to about 56 per cent oxygen in the air at sea-level pressure, nearly three times the normal amount. This limit is approached in the therapeutic use of oxygen in hospitals, where the accepted ceiling is lowered to 40 per cent oxygen, to be on the safe side. For our purposes, however, we will set the upper limit at 400 mm. Hg.

To produce an inspired partial pressure of oxygen of 60 mm. Hg (our minimum) after dilution with water vapor in the respiratory tract, there must be an actual atmospheric partial pressure of oxygen of 107 mm. Hg. If the atmosphere is made up of pure oxygen only, this would represent the total barometric pressure and would be equivalent to just about 2 pounds per square inch (psi). This is about one-seventh the actual barometric pressure of Earth's atmosphere at sea level, which is 14.7 psi. The question then is whether, even though the oxygen supply meets the minimum requirement, man can exist at so low a total barometric pressure.

Apparently, he can. At barometric pressures slightly below this level, gaseous swelling of the body due to the formation of bubbles in the blood has been observed in experiments with animals and human subjects. Carbon dioxide and water vapor, both produced by living tissue in the course of its normal metabolic reactions, are believed to

be the main gases involved in the swelling phenomenon. Thus, the lower limit for oxygen partial
pressure may also serve as a lower limit for total
barometric pressure.

The upper limit of oxygen partial pressure
represents a total barometric pressure of slightly
more than 0.5 atmosphere. However, by diluting
the oxygen with inert gases that have no deleteri-
ous effect on the body, higher barometric pressures
can be reached. Earth's actual atmosphere is di-
luted to a total pressure of 1.0 atmosphere, and
there is no reason at all to think such dilutions can't
reach considerably higher total barometric pressures.

There are only certain other gases that may
be mixed with oxygen without rendering the atmosphere unbreathable, and each has an upper
limit of inspired partial pressure that should not
be exceeded. Symptoms of narcosis (that is, of the
kind of unconscious stupor brought on by narcotics
or anesthetics) have been reported when inspired
partial pressures of nitrogen, argon, krypton, and
xenon—all of which are chemically inert gases—
exceed certain levels. The greater the atomic
weight of the inert gases, the lower the level
required for narcosis. Xenon, the heaviest and
therefore the most narcotic of the stable inert
gases, is narcotic at a pressure of 160 mm. Hg
(0.21 atmosphere). Xenon has, in fact, actually
been used as an anesthetic in surgical operations.

An 80-20 xenon-oxygen mixture, at 1 atmosphere, will produce unconsciousness in 3 to 5 minutes.

Krypton is narcotic at a pressure of 350 mm. Hg (0.5 atmosphere), argon at 1220 mm. Hg (1.6 atmospheres), and nitrogen at 2330 mm. Hg (3 atmospheres). From these figures we can estimate the limiting concentration for the still lighter inert gases, neon and helium (values that have not been determined experimentally). For neon, it is 3900 mm. Hg (5.1 atmospheres) and for helium, it is 61,000 mm. Hg (80 atmospheres). The actual concentration of inert gases in our atmosphere (or any atmosphere likely to be found on an Earth-like planet) is far smaller than these limits. Nitrogen, with an actual partial pressure in Earth's atmosphere of 603 mm. Hg (0.78 atmosphere) comes closest. Its concentration need merely be quadrupled to reach the narcotic level. Argon with a pressure of 6.8 mm. Hg (0.09 atmosphere) is next highest, and its concentration must be increased nearly 200-fold to reach narcotic levels.

Carbon dioxide, although a reasonably inert gas, takes part in certain chemical reactions in living tissue and produces narcosis at considerably lower levels than the inert gases listed above. The upper limit for the carbon dioxide concentration in a breathable atmosphere is 7 mm. Hg. (The actual partial pressure in the Earth's atmosphere is 0.23 mm. Hg, but it is quite possible to find atmospheres with impossibly high concentrations of carbon dioxide. The atmosphere of Venus is very likely a case in point.)

Hydrogen is a special case in that while not poisonous, it is highly inflammable. Only non-combustible mixtures of hydrogen and oxygen could be regarded as acceptable. To be sure, one would never expect to find *both* free hydrogen and free oxygen simultaneously present in a planetary atmosphere. The first lightning stroke would end such an atmosphere, even if it could have been formed in the first place. Methane (a compound of carbon and hydrogen) is similar to hydrogen in being relatively non-toxic but inflammable. Both hydrogen and methane occur in the Earth's atmosphere only in traces. They are continually being formed by the decomposition of organic material on Earth, and are continually being consumed through slow reaction with oxygen.

Our atmosphere contains traces of chemically active gases, other than oxygen itself. These can take part in chemical reactions within the body, inactivating enzymes and distorting the normal path of metabolism. In even low concentrations, then, they are poisonous. Upper limits on the tolerable concentration of such gases as ammonia and carbon monoxide, for instance, are 100 parts per million (ppm); 25 ppm for nitrogen dioxide, 20 ppm for hydrogen sulfide, 5 ppm for sulfur dioxide or hydrogen chloride, and 0.5 ppm for ozone.

These gases occur in our atmosphere in traces that are normally considerably smaller than the upper limits of concentration just listed. All such active gases, by the very fact that they are active, would be expected to react with the free oxygen and water vapor of an atmosphere such as ours and would be reduced to traces quickly (as they are in our atmosphere), even if present originally in larger quantities.

There remains one common gas to be found in planetary atmospheres, and that is water vapor. Water is a special case in many ways. For one thing, it is the only common material with a freezing point within the habitable range of temperature and pressure, so it is the only component of the atmosphere that exists on Earth in the liquid and solid forms, as well as in the gaseous form. Because of its special characteristics, it requires treatment in a separate section.

None of the above gases, other than oxygen and water vapor, is known to be necessary to make up a breathable atmosphere for human beings or, indeed, for any form of animal life. Oxygen is directly necessary; water vapor, indirectly so, to prevent dehydration of the respiratory system; the other gases, for all we know, can be completely absent. To be sure, really prolonged tests on people living in atmospheres containing no inert gases have not yet been carried out, so it cannot be stated categorically that inert gases are unnecessary. (The longest test carried out on human beings in atmospheres containing no inert gases has been of 30 days' duration.)

Some of the inert constituents of the atmosphere are vital to plant life and, therefore, are indirectly necessary to ourselves, even though such gases are not directly involved in respiration. The chief of these is carbon dioxide, which is the necessary source of carbon for the plant world. The normal concentration of carbon dioxide in the Earth's atmosphere is 0.03 per cent, equivalent to a partial pressure of 0.23 mm. Hg. A reasonable minimal value for supporting normal plant life has not been determined, but possibly it would be of the order of 0.05 to 0.10 mm. Hg (about 0.01 per cent).

Some nitrogen is also needed, to supply nitrogen compounds to plants and animals through indirect chemical pathways. Thus, a small fraction of the free nitrogen in the Earth's atmosphere is constantly being converted into the oxides of nitrogen by lightning flashes (and it is estimated that there are over 3 billion lightning strokes per year throughout the world). Some 100 million tons of nitrogen are converted from elemental form into compound form in this fashion; and, without this conversion, plant life on the planet would not be able, in the long run, to obtain adequate supplies of available nitrogen to maintain itself in its present profusion. Atmospheric nitrogen is also converted to usable compounds by bacteria attached to the roots of certain leguminous plants. (Decay processes are, of course, continually restoring molecular nitrogen to the atmosphere.)

The minimum necessary amount of atmospheric nitrogen to support this "nitrogen cycle" has not been determined, but we might place it tentatively at 10 mm. Hg.

We are now in a position to return to the question of the maximum tolerable barometric pressure. Of all possible diluting gases, helium is safest. Helium has been added to oxygen so that the final mixture was 98 per cent helium and 2 per cent oxygen, with a total pressure of 150 psi (about 10 times normal barometric pressure), and such a mixture has been found tolerable for breathing purposes.

Really prolonged exposures have not been studied experimentally, however, and it is not known whether much more than this can be tolerated for long times. As gas density is increased, a point is reached where there would be highly turbulent flow in the air passages of the nose, and the work of breathing would become excessively exhausting. One report states that even at a pressure of 120 psi (8 atmospheres), the turbulence is already so great that one can actually feel eddy currents in the air as it flows through the mouth. Furthermore, only helium can safely dilute oxygen to the point where pressures in excess of 5 atmospheres can be achieved; but that much helium is extremely unlikely to occur in the atmosphere of a habitable planet because such a planet cannot retain helium. By far the most likely diluent in a habitable atmosphere is nitrogen, and nitrogen cannot build up total barometric pressures past the 3-atmosphere mark without reaching narcotic levels. A realistic estimate as to the maximum barometric pressure to be permitted in a breathable atmosphere might, therefore, be 50 psi.

To summarize, then, the atmosphere of a habitable planet must have a barometric pressure between 2 and 50 psi. It must contain oxygen at an inspired partial pressure between 60 and 400 mm. Hg; carbon dioxide at a partial pressure between 0.05 and 7.0 mm. Hg; nitrogen at a partial pressure between 10 and 2330 mm. Hg; and some water vapor. Moderate amounts of xenon, krypton, and argon are permissible, and large amounts of neon and helium, though these gases are most likely to be present in traces only. Inflammable and toxic gases must not be present in more than trace amounts.

Water

Water is, without doubt, one of the most remarkable substances in the Universe and the one most inextricably linked with Earth-type life of all kinds. For one thing, it has remarkable heat-regulating properties. Its high specific heat means that considerable heat is absorbed by water as its temperature rises, and considerable heat must be withdrawn as its temperature falls. Its high heats of fusion and vaporization exaggerate this even further as ice melts to water and water vaporizes to a gas. As a result, water is an excellent heat-storer and can be used with great efficiency as an air conditioner (in the form of perspiration) for the human body and (in the form of oceans) for the Earth itself. The oceans greatly moderate the potential extremes of the Earth's temperature range. Thus, the continental climate of land far removed from the ocean is marked by hotter summers and colder winters than is the oceanic climate of islands and coastal lands at a similar latitude and altitude.

In addition, water expands somewhat when cooled below 39° F. (4° C.) rather than always contracting with falling temperature as most substances do. Its solid form (ice) is less dense than its liquid form, rather than more dense, as is true for most substances. This means that water, cooled to near its freezing point (32° F. or 0° C.), rises and freezes at the surface. The ice formed remains floating on the surface, insulating the water below against further heat loss, so that even the most severe Earthly winter does not freeze completely any sizable body of water. A watery environment of relatively constant temperature is thus preserved through the vicissitudes of Earth's seasons, and in it life forms have originated and evolved.

The incomparable powers of water as a solvent, its high dielectric constant, and its high surface tension, also lend it useful properties in connection with living matter. In particular, its solvent power allows an endless number of reactions vital to life to proceed rapidly, so that life plays out its drama against water as a background. Sixty per cent of the weight of a normally lean man is water, and some forms of simple life are 99 per cent water. It can be said categorically, then, that a habitable planet must have fairly large open bodies of liquid water. This is true even for land life, for not only did land life originate in the ocean, but without oceans there could be no extensive precipitation and hence no salt-free ground water to provide the supplies of fresh water upon which all land life depends.

It is very difficult to determine precisely what ratio of ocean surface area to total planetary surface area is necessary. It is clear that a certain critical total quantity of water is necessary on the surface of a planet before bodies of water can appear. If there were less than this amount, then all of the water would be in the form of water vapor, or of water absorbed on the surface or between the solid particles of the rocky crust. On the other hand, a planet completely covered with water and without permanent dry land could hardly be considered habitable from man's point of view. Since we have already stated that 10 per cent of a planet's surface must be habitable before the planet as a whole can be considered habitable, it seems reasonable to require that the ocean area be less than 90 per cent of the total surface area of the planet.

Extensive open stretches of water imply the continuing presence of water vapor in the atmosphere. This is certainly an additional factor affecting habitability. The uncomfortable effects of high levels of humidity at high temperatures are well known; but there are surely adverse physiological effects due to extremely low levels of water-vapor pressure in the air, too, particularly at the higher temperatures. This latter condition causes very rapid drying of the mucous membranes of the nose, mouth, and throat, and continuous exposure to very low levels of water-vapor pressure might well be intolerable. In a breathable atmosphere, then, water-vapor pressure might vary from relatively small amounts to an estimated maximum of 25 mm. Hg.

Other Requirements

The requirements stated so far for conditions of temperature, light, gravity, atmospheric composition and pressure, and water are probably the major human requisites; yet there are many others. These will be stated briefly.

As we said earlier, *other life forms* must be present, since the very existence of free oxygen in the atmosphere depends on photosynthesis in plants. For a planet to have a breathable atmosphere, it must also, therefore, have the equivalent of indigenous plant life.

The native plant life would not necessarily look anything like Earthly green plants, nor depend on the same complicated series of organic chemical reactions involved in photosynthesis as we know it. All life on the Earth shares certain basic chemical characteristics, but life on other planets could well have somewhat different chemical substructures. For this reason, the native plants might not be edible or palatable to man, and Earthly life forms would have to be introduced as sources of food for human colonists. This would probably be done in any case, since people are likely to prefer and thrive best on familiar foods.

Another requirement is that there must be an absence of unfriendly intelligent beings in prior possession. Man, presumably, can always cope with non-intelligent life forms, however formidable.

Commonly experienced *wind velocities* in otherwise habitable regions must be of tolerable levels. Regions in which wind velocities consistently reach strong gale force (about 50 miles per hour) or higher would not be considered habitable.

Similarly, *dust* normally encountered should be below certain specified levels. It has been suggested that total dust (containing less than 5 per cent free silica—the common ingredient of sand) should not exceed 50 million particles per cubic foot of air, and that high-silica dust (containing more than 50 per cent free silica and, therefore, particularly damaging to the lungs) should not exceed 5 million particles per cubic foot of air. The dust level may depend strongly on the quantity of water on a planet. Water droplets forming on dust nuclei and then falling as rain represent the primary means of removing dust from the atmosphere. Thus, a planet with large oceans and high rainfall should not have a particularly dusty atmosphere, while a planet with small oceans and high winds must be a very dusty place indeed.

The levels of *radioactivity* or ionizing radiation in the atmosphere, whether caused by radioactive materials in the crust or by high-energy particles coming through the atmosphere from outer space, must be of acceptable intensity. The average natural background radiation on the Earth's surface is about 0.003 roentgen-equivalent man (rem) per week. Somewhat more than this can be endured, and the Atomic Energy Commission specifies a steady-state tolerance level of 0.3 rem per week for workers in atomic energy plants. Such a level is too high to be applied to all men on a planet-wide basis, however, because of possible long-term genetic effects. Consequently, it would be desirable to specify dosages from natural background radiation of approximately 0.02 rem per week or less for a planet to be considered habitable.

Other conditions that might render a planet non-habitable would be an excessively high *meteorite-infall rate,* an excessive degree of *vulcanism,* a high frequency of *earthquakes,* and possibly an excessive degree of *electrical activity* (lightning).

Thus we complete a discussion of the principal requirements of human beings with respect to the environmental conditions provided by a planet. We must next move on to a consideration of the variety of planets that actually exist and to a discussion of the astronomical properties they possess. Once that is done, we will be in a position to consider how those properties fit (or do not fit) the environmental requirements of human beings.

/

The Properties
of Planets

General Planetology

About 50 years ago, it was recognized by H. N. Russell and, independently, by E. Hertzsprung that the stars might be grouped into families according to their luminosity and the nature of their spectra.* To illustrate the orderly progression and relationship of these two characteristics, they prepared versions of what is now known as the Hertzsprung-Russell (H-R) diagram. The overwhelming majority of all stars fall into a

---

* The light of a star (or of any source of light) can be spread out into its component colors to produce a "spectrum." The pattern of color, the existence and position of dark areas (representing light absorption) and bright areas (representing light emission), vary according to the chemical composition of the light source, its temperature, its motion, its magnetic properties, and so on. Almost all that we know about stars, aside from position and brightness, has been obtained through careful study of the details of their spectra.

straight, diagonal line on the H-R diagram. This line is the "main sequence," and the stars occupying it are "main sequence stars."

Since the time of this discovery, many general relationships among such factors as mass, luminosity, age, diameter, density, temperature, spectral type, composition, internal conditions, and nuclear reactions have been deduced for stars. Ideas are continuing to be developed about the evolution of stars and about the internal and external physical and chemical changes accompanying the aging processes. Ideas are being developed, too, about the modes of formation of stars, the relationships between members of close double stars, the distribution of types of stars in galactic clusters, and so on.

It is not necessarily true that all current ideas on these matters are correct; many, in fact, are conflicting. The point, however, is that general relationships of some sort do exist. The large luminous bodies of matter in the Universe, called "stars" or "suns," are not individually unique; they are not curious, unrelated specimens that must be studied singly. Stars are, instead, recognized as members of a class of objects with group similarities. Individual stars differ in mass and age, but other observable properties seem to follow inevitably from these primary qualities, plus a very few others such as rate of rotation, propinquity to other massive bodies, and, possibly, original chemical composition. Stars can therefore be considered in groups, and all stars with a particular type of spectrum, for instance, can be expected to have many properties in common.

Astronomers have been able to come to this conclusion because the number of luminous bodies individually detectable in the sky runs into many billions, and because out of these billions no less than 500,000 may be called "well-observed" stars. Consequently, we have available a large population to study and compare; to analyze statistically for number and distribution; to observe spectroscopically; and so on.

Quite a different situation prevails when it comes to the non-self-luminous bodies in the Universe. This class of bodies can be detected only through their ability to reflect light or, very occasionally, through their gravitational effects on nearby stars. The members of this class that are known in some detail are therefore restricted to a few relatively small bodies within our Solar system—bodies including those that are commonly called "planets," "satellites," and "asteroids" (or "planetoids").

Because the sizable non-luminous bodies of the Solar system are so few in number, they have usually been treated as individual objects, each with unique properties, and they have been studied as such. Yet if some current ideas about the formation of stars are substantially correct, there are almost certain to be many more planetary bodies in the Universe than there are stars. It is then to be expected that the class of planetary bodies could be subdivided into families, or classified in a number of ways according to their physical properties and their positions with respect to nearby stars. From this point of view, then, the planets of the Solar system might be regarded not as unique specimens but as members of large families of such objects, in which the relationships between the definable physical characteristics would be found to follow certain general laws of nature. It is only because we have so few to study that these relationships have not become perfectly obvious.

The word "planetology" has been used in the past to mean the "study and interpretation of surface markings of planets and satellites." A broader term is required to cover not only the surface markings, but all the physical properties of non-self-luminous bodies, whether they are a part of our own Solar system or are orbiting about some other star. For this purpose, the term "general planetology" is proposed, and is defined here as "a branch of astronomy that deals with the study and interpretation of the physical and chemical properties of planets." In this context, planets will then be defined as "massive aggregates of matter that are not large enough to sustain thermonuclear reactions in their interiors." (If they *were* large enough to sustain such reactions, they would become self-luminous and would then be classified among the stars.)

At present, it is obvious that our knowledge of the underlying laws of general planetology must be far from complete. For one thing, some of the properties of the planets of the Solar system (on which we depend for our starting point) are known only approximately; our information is not reliable enough to be used as part of the firm foundation of the study.

The density of Mercury, for example, has been reported by some observers to be as low as 3.7 grams per cubic centimeter (g/cm^3), and by others to be as high as 6.2 g/cm^3. Numerous intermediate values have also been given. This variation in reported density is a consequence of the extreme difficulty of measuring with precision either the mass or the diameter of Mercury. (If the diameter were accurately known, the volume of Mercury could be easily and precisely calculated. If the mass were also accurately known, then that mass divided by the volume would give the density. However, Mercury is too small and too difficult to observe near the glowing Sun for its diameter to be easily determined, and the fact that it lacks a satellite deprives us of the surest method of determining its mass—which would be to measure the period of revolution of such a satellite about Mercury and its distance from the planet.)

The physical dimensions and densities of Uranus and Neptune are also known only approximately, while scarcely anything at all is known about Pluto. In addition, our current state of knowledge concerning the properties and behavior of ordinary matter under extreme conditions of pressure (as must exist in the interiors of planets) is still quite rudimentary. It cannot yet be said that we have anything but a fragmentary picture of the causes of mountain-building processes, earthquakes, and volcanoes, or of the structure of the Earth's crust, its mantle, or its core.

Finally, there are many areas of study in which the complete working out of all the effects that could take place is so extraordinarily difficult and complex when using present techniques that it is often necessary to simplify the problems greatly to handle them at all. A case in point is the current state of Earth meteorology. Great strides are being made in the understanding of winds, storms, precipitation, air circulation patterns, and other atmospheric phenomena; but many questions still remain unsolved (the cause of the ice ages, for instance, to take a spectacular example), and most weather predictions must still be made on a largely empirical basis. How much more difficult would it be, then, to elucidate planetary meteorology in a completely general manner, taking into account every possible combination of atmospheric composition, planetary mass, surface gravity, rate of rotation, land-sea ratio, tilt of the equatorial plane, and so on?

Despite all this, there remain certain overriding or dominant astronomical factors that permit estimations of the more general aspects, at least, of planetary meteorology. A knowledge of the relative universal abundances of the chemical elements, coupled with an understanding of the chemical and physical properties of the most abundant elements and compounds, permits us to make deductions as to the constituents that are likely to be part of the original makeup of a planetary atmosphere. Then, too, the conditions necessary for the loss or retention of various gaseous atmospheric constituents can be estimated, and, in this way, certain subclassifications of planets can be derived according to the type of atmosphere they eventually retain.

The main objective of general planetology, in common with all science, is a fuller understanding of the Universe in which we live. Some subsidiary objectives are to define the general characteristics of planetary systems; to gain a clearer understanding of the characteristics of habitable planets and to obtain a more definitive estimate of the number of habitable planets in our Galaxy; to indicate which of the stars in the neighborhood of the Sun would be most likely to possess habitable planets and what the probabilities of such possession might be; and, finally, to obtain a better understanding of our own planet and an appreciation of the combination of factors that makes the Earth a comfortable place to live.

With all that in mind, we begin by turning to some of the general properties of the planets of our own Solar system.

The Upper Limit of Planetary Mass

It would seem natural to begin a discussion of the properties of massive aggregations of matter in the Universe with a description of the modes of formation of the stars and planets. This subject is a controversial one, but the theories most commonly accepted nowadays picture the stars as originating by the slow gathering together ("accretion") of dust and gas within clouds of cosmic matter, and planets as formed in a similar manner from the leftover matter surrounding the newly formed stars. The "accretion hypothesis" seems to account quite satisfactorily for most of the presently observed properties of the bodies of the Solar system in a way that no other hypotheses do. Thus, without going into the minute details (about which there is much argument), we will assume here that the accretion hypothesis is correct and pass on to the Solar system as it now exists.

By far the most important basic property of large aggregations of matter is mass. The very distinction between stars on the one hand and planets on the other can be made to rest almost exclusively on mass. If the mass of a body is large enough, internal pressures within it will raise the temperature to the point of triggering and sustaining thermonuclear reactions in its interior. Such bodies will be stars. If the mass is not large enough to support such thermonuclear reactions, the body is non-self-luminous and is therefore a planet.

All of the stars of which we have any observational knowledge are in the mass range from 0.04 to about 60 times the mass of our Sun. What's more, the vast majority of stars that we can observe fall into a much narrower mass range, between 0.2 and 5 times the mass of our Sun. This limited mass range of stars was surprising when first discovered; but astronomer A. S. Eddington explained such a limitation on the basis of the balance between gravitational force tending to contract a large body and radiation pressure tending to expand it.

Eddington defined a star as any aggregation of matter with a mass between 10^{32} and 10^{37} grams; that is, from 0.05 to 5000 times the mass of our Sun. He held that a body of less than 10^{32} grams in mass could not remain self-luminous as a star and that an aggregation of matter exceeding 10^{37} grams in mass would blow itself apart by the pressure of its own radiation.

If we reach below the minimum mass level for stars, we reach the region of non-stars, or planets. Clearly, there is some upper limit to the mass of a planet and some lower limit to the mass of a star, although the exact value of mass at which this transition takes place is not known at present. Indications are, however, that the transition region lies somewhere between 1000 and 10,000 times the mass of the Earth. This may also be expressed as between 0.003 and 0.03 times the mass of our Sun, or as between 6×10^{30} and 6×10^{31} grams—the region just below Eddington's minimum.

Unfortunately, no bodies with masses lying in this transition region are well known. Of the bodies of our Solar system, the largest next to the Sun itself is Jupiter, which has a mass 317 times that of the Earth. It would have to be over 3 times as massive as that to fall into even the lower limits of the transition region. Three other planets of the Solar system are more massive than the Earth. Saturn has a mass 95 times that of the Earth, Neptune is 17.5 times the mass of the Earth, and Uranus is 14.5 times the mass of the Earth. All the other non-self-luminous bodies of the Solar system are less massive than Earth.

It is possible that increasing knowledge will not completely sharpen the value of mass at transition: that the dividing line between the largest planet and the smallest star may be inexact by its very nature because minor effects are produced by factors other than mass. If so, there may even be overlapping, so that some large planets will prove to be slightly more massive than some small stars.

Thus, if the rate of matter accumulation by a growing planet depends not only on its mass at the moment, but also on the local density of unaccreted matter and the relative velocity of that matter, then it follows that two objects ending up with the same mass may have had quite different rates of growth. One planet, surrounded by a low density of unaccreted matter, would have had a slow rate of growth, permitting a longer time to radiate away the kinetic energy resulting from the accretion of moving particles falling into the planet. This might result in central temperatures not quite high enough to trigger off the thermonuclear reactions necessary for a star. If it had been surrounded by a higher density of unaccreted matter, it might have grown more rapidly, reaching the same mass but having less time to get rid of heat. It would then have reached the trigger point and become a star.

A borderline class of objects might conceivably exist, too: objects that are just able to trigger off thermonuclear reactions but lose this ability on the expansion that follows their rise in temperature. On consequent cooling and contraction, internal temperature would rise and trigger the thermonuclear reactions again. Such a body would then oscillate or pulsate weakly on the border line between planet and star.

The growth rate of a planet would be strongly affected by the presence of larger objects in the same system. These larger objects, by rapidly adding to their own mass thanks to their intense gravitational field, would quickly reduce the mean density of the matter available for the growth of the planet. Usually, it is the central star that is the strongest competitor for growth material.

The farther away the orbit of a planet from the large central star (all other things being equal), the less the two tend to compete for growth material. Also because the star's gravitational field weakens with distance, the local material in the neighborhood of the distant planet possesses a lower velocity than similar material in the neighborhood of a planet close to the star. Both of these factors—the lessened competition between planet and star and the lessened velocity of passing material—increase the rate at which a planet can capture material and grow. A planet distant from the star, therefore, may well attain greater mass than a planet orbiting nearer the star, other things being equal. In our Solar system, this rule is evident, for the planets more distant from the Sun (Jupiter, Saturn, Uranus, and Neptune) are considerably more massive than those relatively close to the Sun (Mars, Earth, Venus, and Mercury).

The Lower Limit of Planetary Mass

As for the lower mass limit for planets, that is not so clearly defined. There is no change with decreasing mass that is quite so obvious as the change from non-self-luminosity to luminosity, which, with increasing mass, converts a planet to a star. It is difficult to decide, then, at what level of smallness a body is not a small planet but merely a large meteoroid. By studying the relationship of the mass of non-self-luminous bodies to other characteristics, however, we may find some criterion that will allow us to define the word "planet" with reasonable precision, by setting a lower as well as an upper limit of mass.

To begin with, let's consider those bodies in the Solar system that are comparatively close to the Sun and that have properties like those of the Earth in many respects. These are the "terrestrial bodies." If we compare the mass and density of those terrestrial bodies for which reliable data are available, an interesting and logical pattern emerges. We find, in general, that the more massive the body, the more it is compressed, as a whole, by its own gravitational force and, therefore, the greater is its density.

The most massive of the terrestrial bodies is the Earth itself, and it is also the densest, 5.52 g/cm^3. Venus, which possesses a mass 0.816 times that of the Earth is almost as dense, 5.32 g/cm^3. Mars is much smaller, only 0.1077 times as massive as the Earth, and it has a density of 4.0 g/cm^3; while the Moon, which is smaller still, only 0.01229 times as massive as the Earth, has a density of only 3.34 g/cm^3. We can carry this to an extreme by considering Earth's surface rocks. They are pulled down by Earth's gravity but are not compressed by any overlying strata of rocks. The simple pull of an outside gravitational field is small enough to be ignored, and Earth's surface rocks can be considered as bodies of zero mass. Their average density is 2.8 g/cm^3.

Given the mass and density of a body, one can easily calculate the volume of that body, and from that one can further calculate its radius (the distance from the center to the surface). The densities of such bodies are then found to vary quite consistently with their radii, so that one can easily predict the density of a terrestrial body from its size alone. To be sure, the apparent density of Venus falls a trifle below the value that one would expect from the relationship. However, the actual radius of Venus is uncertain. What we see as the rim of Venus' structure is the outer limit of its cloud layer, and we are not certain what the depth of the atmosphere below the cloud layer is. The rocky sphere beneath the clouds must have a smaller radius than the one we observe, and if that smaller radius is taken, then there is little doubt that the density of Venus will be found to fall into place.

Mercury is a special case. As we explained above, there is great difficulty in determining its density. The wide range of figures given for the density includes the figure that we would expect to follow from the relationship of radius and density. However, the most likely values seem to be distinctly higher than that figure. If this is not the result of errors in observation but is actually so, the irregularity may be associated with Mercury's proximity to the Sun. Since Mercury apparently keeps one face perpetually toward the Sun, the surface rock temperature on the sunny face may reach about 1300° F. (700° C.). Temperatures of this order of magnitude, accompanied by high vacuum, can bring about the loss of water of crystallization from some common rock minerals, which will then become capable of more condensed packing, so that their density can increase. It is also possible that the rocky material of which Mercury is composed lost gas and water through solar heat even before its aggregation into a planet.

Data on terrestrial-type bodies in the Solar system beyond the orbit of Mars, such as the asteroids and the large satellites of Jupiter and Saturn, are not sufficiently reliable for inclusion in this mass-density relationship. Still, what we have with the material on hand is an example of the manner in which planetary properties follow regular rules so that planets need not be considered as unique, unrelated specimens.

Furthermore, from this regular relationship, other characteristics can be deduced. As has been said, a regular relationship between mass and density implies a regular relationship between a planet's volume and its radius with density. From the mass and radius, one can calculate the surface gravity and the escape velocity,* each of which then also proves to have a regular relationship to density.

However, let us follow the line of argument in still another direction. The more massive a body of a given composition, the more easily will its parts change shape or deform under its own gravitational forces. (A more massive body is more dense because the compression of its inner layers increases, and this compression is a form of deformation.)

Small chunks of rock, in other words, can exist in almost any conceivable shape; but a large accumulation of matter is deformed under its own gravitational forces and cannot maintain an irregular shape even if one existed to begin with. On a large mass of matter, loose particles roll "downhill" under the force of gravity. Toward the center of the mass, where the pressures of overlying rock are greatest, even the strongest rigid materials flow like liquids until an equilibrium has been established and every particle is as close to the center of gravity as possible.

The situation in which every particle of a planetary body is as close to the center of gravity

_____

* An object thrown upward attains a certain height. The greater the initial velocity of the thrown object, the greater the height attained. If the velocity is high enough, then the height is, theoretically, infinite, and the object never returns. The velocity at which this occurs is the "escape velocity." For the Earth, the escape velocity is 6.98 miles per second.

as possible is attained when the body takes on a spherical shape. (There would be minor departures from the spherical shape because of the effects of rotation or because of differences in density among the surface material; but, on the whole, we could call the equilibrium shape of a sizable aggregate of matter a sphere, usually as a very close approximation.)

This means, then, that we can divide non-self-luminous bodies into two classes: those small enough to be able to maintain highly irregular shapes, and those large enough to be forced into spherical shape. We can go on to apply the word "planet" only to the latter, and so we have our lower limit of planetary mass: It is the lowest mass at which an irregular shape is impossible and a spherical shape is enforced.

The greatest possible mass of a body capable of preserving a highly irregular shape seems to be about 0.00001 that of the Earth. Such a body would have a radius of 125 miles and would be about the size of a large asteroid. For materials with particularly high yield strengths, the transition from irregular to more nearly spherical shapes may take place at slightly higher values of mass, but no known material has a yield strength high enough to maintain an irregular shape when it has a mass as much as 0.0001 times that of the Earth and a radius of about 275 miles.

(Before leaving them permanently, we might pause to consider some of the properties of small rock masses. A man can jump off any body having a mass less than about 7×10^{17} grams [and a radius of less than 2.4 miles] if we assume that he can jump with an initial velocity of 16 feet per second. A man can throw an object such as a baseball completely away from any body having a mass less than about 2×10^{20} grams [and a radius of about 16 miles], assuming that he can launch it at about 110 feet per second. A rifle bullet can be shot away from any body having a mass less than about 3×10^{24} grams [and a radius of about 400 miles] assuming a muzzle velocity of 2700 feet per second. In fact, people exploring small asteroidal bodies in this size range will have to be careful about throwing or shooting, for objects launched horizontally at velocities less than escape velocity might go into orbit about the asteroid. They would remain in such an orbit indefinitely and could constitute a hazard to personnel each time they skimmed back in to make their closest approach.)

Well, then, we will apply the word "planet" to any massive body between the mass limits 0.00001 to 10,000 times the mass of the Earth. What's more, we will not limit the term to those bodies of our Solar system that are revolving about the Sun. We will apply it to bodies in this mass range whether they are revolving about the Sun, or about another planet, or about some distant star, or whether they are isolated in space.

Within the Solar system, there are thousands

upon thousands of small objects that do not fall within the planetary region of mass by our definition. There are, however, about two dozen bodies that do. Of these 9 are those bodies ordinarily called planets: Mercury, Venus, Earth, Mars, Jupiter, Saturn, Uranus, Neptune, and Pluto. In addition, there are 12 bodies, ordinarily called satellites, which nevertheless fall into the planetary mass range. These include Earth's satellite, the Moon; Jupiter's four largest satellites, Io, Europa, Ganymede, and Callisto; Saturn's six largest satellites, Tethys, Dione, Rhea, Titan, Hyperion, and Iapetus; and Neptune's larger satellite, Triton. Finally, there are the three largest known asteroids: Ceres, Pallas, and Vesta. Actually, the masses of the satellites (except the Moon) and of the asteroids are not known with much accuracy, so some of those named above may have masses below our arbitrary lower limit, while a few others are borderline cases: Saturn's satellites, Mimas, Enceladus, and Phoebe; Uranus' satellites, Ariel, Umbriel, Titania, and Oberon; and Neptune's smaller satellite, Nereid.

Gas Capture and Retention

As we explained earlier, there is a clear relationship between the mass and the density of the terrestrial bodies of the Solar system, and the largest body we considered in connection with this relationship was the Earth itself. There are, however, four bodies in the Solar system that are distinctly more massive than the Earth. These are Jupiter, Saturn, Uranus, and Neptune. How do these fit into the relationship?

The answer is that they fit the relationship very poorly; in fact, not at all. The four massive planets, though more massive than the Earth by far, are nevertheless less dense than the Earth. They are less dense than any of the terrestrial bodies; less dense than the Moon; less dense, even, than Earth's surface rocks. The density of Neptune is about 2 g/cm³ and that of Uranus is about 1.5 g/cm³. Jupiter has a density of 1.34 g/cm³ and Saturn has one of 0.69 g/cm³.

The reason for the difference is not hard to see. The terrestrial bodies are rocky spheres without a sufficiently large gravitational field to attract much of an atmosphere if, indeed, they accumulate any measurable gaseous envelope at all. The situation is different for bodies that attain a mass about 3 times that of the Earth or more. Their gravitational field is high enough to accumulate considerable atmosphere. The gases of such an atmosphere are low in density compared with the rocky sphere itself, and their presence greatly reduces the average density of the planet as a whole. (Indeed, as was mentioned earlier, Venus' comparatively small atmosphere was enough to lower the over-all density below the expected value. The effect becomes more extreme as the atmosphere grows more extensive.)

Up to a certain limit, increasing mass implies a larger and larger accumulation of gas and a lower and lower over-all density. Thus, Saturn, with a mass 6 times as great as that of either Uranus or Neptune, has a distinctly lower density than either of those two bodies. Saturn, however, is near the limit of this effect. As masses grow still higher, the gravitational field becomes high enough to compress the atmosphere to densities so high that the effect of gathering gas no longer serves to continue reducing over-all density. Thus Jupiter, with a mass 3.3 times that of Saturn, is also somewhat denser.

The existence of extensive atmospheres on these massive planets effectively hides the nature of their internal composition. Until more is learned about the behavior of matter under extreme conditions of pressure, reliable quantitative estimates of the internal composition of these massive bodies cannot be made, although if some assumptions are made regarding their modes of formation, some limits on their internal constitutions may be inferred. Much depends on the temperature conditions assumed at the time of formation and on the present temperatures in the upper levels of the atmosphere, for it is these upper temperatures that dictate (in part) the rate at which the gases of the atmosphere can escape from the neighborhood of the planet.

A consideration of this tendency of atmospheric gases to escape from the planet shows

plainly enough that the size and chemical nature
of a planet's atmosphere does not depend on the
planet's mass alone. One other very important
factor, at least, is its temperature.

A number of theories on the escape of planetary atmospheres have been developed in recent
years. Unfortunately, the more complex of these
theories require a detailed prior knowledge of the
variation of temperature with changes in altitude
within the atmosphere. In addition, there must be
a prior knowledge of the atmospheric composition. Unfortunately, we do not have such knowledge in detail for any planet but the Earth, and
theories based on this knowledge cannot be applied conveniently to generalized cases.

A more useful, though rough, yardstick to
measure the escape of atmospheric gases was first
derived by Sir James Jeans in 1916. This relates
the time of escape of a particular gas to such
properties as the radius of the planet, its surface
gravity, and, in particular, the average velocity of
the molecules of gas in the atmosphere. (To be
precise, the "average velocity" of these molecules
should be referred to as the "root-mean-square
velocity," or "rms velocity," a phrase taken from
the arithmetical procedure used to obtain this
particular type of average.)

The higher the rms velocity of the gas molecules, the more difficult it is for the planet to accumulate an atmosphere in the first place, and the
more readily will the planet lose an atmosphere

already accumulated. For any given gas, the rms velocity of its molecules rises with temperature. Thus, a particular planet can more easily accumulate (or retain) an atmosphere at low temperature and more easily lose it (or fail to accumulate it in the first place) at high temperature.

Specifically, if the rms velocity of the gas in question at a particular temperature equals the escape velocity of a planet, many of the individual gas molecules will, at some particular moment, be bound to have velocities 2 or 3 times as high as the over-all average. Some of these molecules are bound to be moving in an upward direction. Consequently, the gas will escape rapidly and permanently from the neighborhood of the planet. Even if the rms velocity is only one-half the planetary escape velocity, there will be enough molecules moving faster than escape velocity to ensure rapid loss of the atmosphere. If the rms velocity of the molecules is one-third the escape velocity, the planet's atmosphere will last a few weeks; if it is one-fourth the escape velocity, it will last several thousand years; and if it is one-fifth the escape velocity, it will last about 100 million years. Finally, if the rms velocity of the molecules is one-sixth the escape velocity or less, then the atmosphere is essentially permanent. No significant quantity will be lost over the life of the planet, provided its surface temperature undergoes no considerable rise at any period. For a planet to retain its atmosphere, then, the rms velocity of the molecules of its gaseous envelope must be no more than about one-sixth to one-fifth the escape velocity. The rms velocities of oxygen and nitrogen in the Earth's atmosphere lie well within that limit, of course, and the Earth has no trouble retaining its atmosphere indefinitely.

There are conditions under which a planet can capture or retain an atmosphere even when the rms velocity of the gas molecules is, initially, too high for the gas to be retained. The paradox is explained by the fact that no matter what the rate of atmospheric loss, the atmosphere will continue to grow if gas is arriving at a still greater rate. As the atmosphere accumulates, the planetary mass will increase and with it the escape velocity. Eventually the rms velocity of the molecules in the atmosphere will be sufficiently small in comparison to the increasing escape velocity for the rate of loss to fade off to negligible values. The rate of net atmospheric accumulation will speed up, and the gas capturing process will accelerate until it is terminated by a lack of surrounding matter.

In the accumulation of such an atmosphere, the only gases that really count are hydrogen and helium. It is estimated that hydrogen makes up 90 per cent of the material of the Universe and helium, 9 per cent. All other substances make up 1 per cent or less of the Universe, and, in any massive growth process, their contribution can be ignored.

The rms velocity of a gas at a particular temperature varies according to the mass of the atoms or molecules composing it (its "atomic weight" or "molecular weight"). The lower the molecular weight, the higher the rms velocity at a given temperature. The gas with the lowest known molecular weight is hydrogen, and hydrogen molecules (each made up of a pair of hydrogen atoms) have a higher rms velocity at a particular temperature than do any other molecules. It follows that hydrogen is the most difficult gas for a planet to capture or to retain if captured.

Helium is composed of single atoms, each of which is twice as heavy as the hydrogen molecule. Helium therefore has a lower rms velocity than hydrogen and is correspondingly easier to capture. It is the more easily captured helium that in all likelihood starts the "snowballing" process of gas capture. Once a certain amount of helium has been captured, the increased mass and escape velocity allow even hydrogen capture to proceed rapidly, and a giant planet such as Jupiter or Saturn is the end result.

This "snowballing" effect can produce great changes with a small initial change in mass. Terrestrial planets consist almost entirely of combinations of oxygen with the elements silicon, aluminum, iron, calcium, magnesium, sodium, and potassium. The resulting rocky material is thus made up of elements other than hydrogen and helium and, consequently, out of 1 per cent of the Uni-

verse as a whole. Once helium and hydrogen begin
to be collected, the solid core of a terrestrial planet
may be quickly swamped. A solid core about 3
times the mass of the Earth represents the approximate minimum mass at which "snowballing" takes
place and at which sizable atmospheres are accumulated. By the time a massive planet has
accumulated gas and dust and has grown to 6.5
Earth masses, it might be half solid and half
gaseous; after having grown to 14 Earth masses, it
might be only one-quarter solid and three-quarters
gaseous.

Do Jupiter and the other outer planets possess
rocky cores that are much more massive than the
Earth? The obscuring atmosphere of those planets
makes it difficult to say, although it would certainly seem highly probable that they do. In addition, however, there is a temperature effect that
also works in favor of the outer planets as far as
"snowballing" is concerned.

What is important in this respect is the temperature of the very rarefied upper region of a
planetary atmosphere, where the gas density is so
low that individual atoms or molecules may travel
several miles before finding other atoms or molecules with which to collide. (The length of non-
colliding travel is called the "mean free path.")
This rarefied region of the atmosphere is usually
termed the "exosphere." In the exosphere, a very
fast molecule moving vertically away from the
planet has an excellent chance of moving indefi-

nitely away from the planet without being bounced downward again by collision. It is from the exosphere, then, that an atmosphere is lost, and the rate at which it is lost depends on the temperature there, rather than on the temperature of the atmosphere at sea level. The higher the exosphere temperature, the greater the rms velocity of the atoms and molecules there, and the more rapid the escape (or, what is much the same thing, the less rapid the accumulation).

For the Earth, the critical escape layer is apparently at an altitude of about 600 kilometers (375 miles). The temperature at this altitude is quite variable because of the changing intensity of solar radiation, but it is quite high, thanks to the far-ultraviolet radiation of the Sun. Estimated temperature falls within the range of 1000° K. to 2000° K.*

The temperatures at the critical escape level in the exospheres of other planets are even less certain than is the value for the Earth. However, we can start with a temperature of 2000° K. for the Earth's exosphere and assume, for example, as a crude estimate, that the exosphere temperatures of other planets vary according to the distance of their closest approach to the Sun.

Thus, Mercury, which approaches to within 28.5 million miles of the Sun, is closer to the Sun

---

* These are absolute temperatures (degrees Kelvin, or °K.), usually used for the higher values. If 273 is subtracted from an absolute temperature value, the equivalent temperature on the Celsius scale is obtained.

than the Earth ever is, by a factor of 3.3. Mercury's exosphere temperature,* then, would be expected to be considerably higher than that of the Earth. On the other hand, Jupiter never approaches closer than 460 million miles to the Sun. It is farther away than we are by a factor of 5.2. Its exosphere temperature should be a great deal lower than ours.

Jupiter and the planets beyond tend to undergo the "snowballing" effect, then, for two reasons. First, they very likely have solid cores considerably more massive than the Earth. After all, they are far from the Sun and suffered less from competition for growth material with the Sun (see page 52). Second, they have lower exosphere temperatures and for that reason can more easily capture and retain helium and hydrogen. Of these outer planets, Jupiter, the nearest to the Sun, is the most massive; Saturn, farther out, is less massive; and Uranus and Neptune, still farther out, are still less massive. This could be a reflection of the fact that in the early days of the Solar system, the density of growth material decreased with distance from the Sun.

Light Atmospheres

Of all the planets in the Solar system, then, only Jupiter, Saturn, Uranus, and Neptune had

---

* Mercury does not have an atmosphere in the ordinary sense, but a very thin vapor may cling to it, enough to give it what may fairly be called an exosphere. The same is true of the Moon.

cores massive enough, or exospheres cool enough, or both, to initiate the "snowballing" effect; and those four are the only planets with massive atmospheres. They are sometimes called the "gas giants."

The remaining planets had cores insufficiently massive, or exospheres too hot, or both, to be able to capture helium or hydrogen and thus never developed a massive atmosphere. To be sure, oxygen and nitrogen are relatively heavy gases when compared with hydrogen and helium. The oxygen molecule (composed of two oxygen atoms) is 16 times as heavy as the hydrogen molecule; the nitrogen molecule (composed of two nitrogen atoms) is 14 times as heavy. The Earth is massive enough and its exosphere temperature is low enough for it to have been able to capture these gases if they had existed in any quantity in the growth material. They did not, however. Any substance other than hydrogen or helium existed in comparatively small quantities, too small to represent easily captured material.

It is doubtful, then, whether the atmosphere of the Earth (or any atmosphere made up of gases other than hydrogen or helium) was captured from surrounding space. It is quite likely that the Earth, and all the planets other than the gas giants, were first formed without an atmosphere, and that it was volcanic action that then provided the primary ingredients from which the present atmosphere (and the oceans as well) developed—these primary ingredients having been originally bound

physically or chemically within the rocky structure of the planet itself. It is even possible that
some atmospheric constituents that normally
would escape from a particular planet may be
supplied by volcanic action as quickly as they
escape so that a certain concentration is always
present in the atmosphere. Such constituents may
also be supplied by photolysis (the breakdown of
more complicated compounds into simple gases
through the action of sunlight), or by radioactive
breakdown. Thus, although the Earth is incapable
of retaining helium in its atmosphere, small traces
of the gas are present in the atmosphere and will
continue to be present because the gas is continually leaking out of the soil as a result of being
formed there in the course of the slow radioactive
breakdown of thorium and uranium compounds.

But let us return to the major products of volcanic action. According to recent studies, water is
the chief component of volcanic gases at the
Earth's surface, generally constituting more than
75 per cent, by volume, of all gases collected at
volcanic vents. The predominance of water makes
it probable that in the course of the geologic ages,
all the water on the Earth's surface has been produced by volcanoes. (It should also be mentioned
that volcanic action was, very likely, more intense
in the early periods of Earth's history.)

Other volcanic gases, too, have accumulated
over the several billion years since the Earth was
formed, and a number of important physical and

chemical changes have taken place as a result. In the presence of water, carbon dioxide (the most common of the volcanic gases next to water) was removed from the atmosphere and converted into carbonate rocks. Other water-soluble, chemically active gases produced by volcanic action (such as ammonia, hydrogen sulfide, and sulfur dioxide) were also dissolved and converted into various minerals. The nitrogen and argon produced by volcanic action were retained unchanged in the atmosphere. The early atmosphere, then, consisted of nitrogen, argon, and whatever other gases remained in equilibrium with their dissolved forms in the gathering ocean.

Once life developed and photosynthesis began, additional carbon dioxide was removed from the atmosphere by the action of the green plants and was replaced by oxygen. In the presence of a growing excess of free oxygen, any carbon monoxide or methane that had been produced by volcanic action was oxidized to carbon dioxide, which was also thrown into the photosynthetic furnace. Such hydrogen as was produced either escaped, or combined with oxygen to form water. Little by little, then, through reactions involving both life and non-life, Earth's atmosphere evolved its present mass and chemical composition.

If vulcanism is accepted as the primary mechanism for the production of atmospheres of planets that are not massive enough to capture hydrogen and helium in large quantities, then an understanding of the relationship between degree of volcanic activity and planetary mass is essential for an understanding of the general course of atmospheric evolution. Actually our knowledge of the natural forces responsible for volcanic activity, earthquakes, and mountain formation on Earth is still quite incomplete; thus it is difficult at this time to specify general relationships between planetary mass and volcanic action, and, therefore, between planetary mass and the fine details of any light atmosphere that may be formed.

One view is that earthquakes and volcanic heat result from the mechanical energy associated with the distortion of the crust. Distortions of the crust would be produced either by a general increase in the temperature of the interior of a planet with subsequent expansion, or by a general cooling with subsequent contraction. On the whole, the higher the internal temperatures, the greater the changes either way and the more pronounced the volcanic action.

The high interior temperatures of the Earth are due, apparently, both to gravitational compression and to the accumulation of heat released by radioactive materials such as uranium, thorium, and potassium-40. (The last is a comparatively rare isotope of the common element potassium.) Radioactivity is apparently a more important contributor to internal heat than is gravitational compression.

A planet smaller than the Earth would tend to accumulate less heat through gravitational compression during the period of formation. Then, too, what heat was formed, either through gravitational compression or through radioactivity, would escape more readily from a smaller planet than from a larger one.

To see the reason for the more rapid heat loss, consider that the ratio of surface area to volume increases as an object of a particular shape decreases in size. A small spherical planet has more surface area for its mass than a large spherical planet. (If a sphere is reduced to one-eighth its former mass, its surface area is reduced by only one-fourth. By reducing the sphere in this fashion, the ratio of surface to mass has doubled.) Since heat is lost through the surface area, then, a small sphere would lose its heat at a more rapid rate than a large sphere, for it would have more surface area (in proportion) through which to lose the heat.

Small planets might also tend to have less concentration of metals toward the center; and since metals are good conductors of heat, if they are spread more evenly through the planet, the planet as a whole (and especially its outer regions) is a better heat conductor. As a result, internal heat is lost more rapidly.

In short, there is every reason to think that small planets have lower internal temperatures than large ones and, consequently, have less volcanic activity and less gas formation than large planets. It is not surprising, then, that Mars, considerably smaller than the Earth, also has a considerably thinner atmosphere.

The atmospheres of Earth and Mars differ not merely in total mass and in chemical composition. (The Martian atmosphere lacks more than a trace of oxygen.) They vary in another and more subtle fashion.

The manner in which the density of a planetary atmosphere varies with altitude above the surface depends, to a great extent, on the strength of a planet's gravitational field. A large planet with a high surface gravity compresses the lower layers of its own atmosphere under the weight of the upper layers. The density of the atmosphere therefore changes markedly with changes in altitude, dropping off rapidly with height. A small planet, with a low surface gravity, has a smaller compressing effect on its atmosphere, which therefore shows a more gradual decrease in density with increased altitude.

On Earth, for example, the atmospheric density drops by a factor of about two for each 17,000 feet of ascent above the surface. On a larger planet, having the same atmospheric molecular weight and temperature conditions, this halving of density would take place in altitude intervals of less than 17,000 feet. On a smaller planet, the density-halving would take place in altitude intervals of more than 17,000 feet. A large terrestrial planet might therefore be said to have a "hard" atmosphere, and a small planet a "soft" atmosphere. For example, Mars has a softer atmosphere than Earth has.

This relationship between size and density change in the atmosphere has a great bearing on the ease of entry of a space vehicle into a planet's atmosphere. Entry into the atmosphere of a small planet such as Mars can be achieved with less rigid restriction on angle of entry and with lower accelerations imposed on the passengers than entry into the atmosphere of larger planets such as the Earth.

To summarize, then, we see that planets can fall into three atmospheric classes:

1. Planets without measurable atmospheres. Such planets have a gravitational field too small to retain even those common gases with the most massive molecules. Perhaps most planets fall into this class, since, in all likelihood, there are more small planets than large ones, and a small planet has a small gravitational field. The best-known examples include Mercury, with its unusually high exosphere temperature, and the Moon, which has a lower exosphere temperature than Mercury has, but also possesses a distinctly smaller mass and, consequently, a lower escape velocity.

2. Planets with light atmospheres. These have gravitational fields too small or exosphere temperatures too high to capture hydrogen or helium and consequently do not undergo the "snowballing" effect and do not collect massive atmospheres. However, their gravitational fields are large

enough to retain moderately heavy gases such as
nitrogen, oxygen, and carbon dioxide, where these
are produced by volcanic action. The one habitable planet of the Solar system, the Earth, falls into
this class. Venus is another member. Mars is too
small to be an efficient gas retainer or to have
the extent of volcanic action probably found on
more massive planets such as Earth and Venus.
Nevertheless, it too retains an atmosphere (thinner
than that of the Earth or Venus), aided by its
exosphere temperature, which is lower than that
of Earth or Venus.

3. Planets with massive atmospheres. These
have exosphere temperatures low enough and
rocky cores massive enough to allow them to initiate the "snowballing" effect. The atmospheres
collected are composed largely of helium and hydrogen, and the four known examples of this class
are Jupiter, Saturn, Uranus, and Neptune.

The Oblateness of Rotating Planets

Most of the discussion up to this point has
been concerned with the properties of massive
bodies that are not rotating rapidly. The rate of
rotation, however, is an important property of a
planetary object, affecting its shape, surface gravity, and habitability. In a consideration of general
planetary properties, the effects of rotation cannot
be ignored.

The shape of a rotating body isolated in space

depends on its rate of rotation, its average density, and the distribution of mass within the body. A rotating body experiences a centrifugal effect which produces an upward pull countering part of the downward pull of gravity. The strength of this centrifugal effect on a particular portion of the planet's surface varies according to the velocity with which that portion moves. In the case of a planet rotating about an axis, the surface does not move at all at the poles, but moves with greater and greater velocity as the distance from the poles increases, reaching a maximum, of course, at the equator. Thus, on Earth, the velocity of rotation for an object on the surface at Anchorage, Alaska, or Leningrad, U.S.S.R. (60° North Latitude), is 520 miles per hour. At Peking, China, or Philadelphia, Pennsylvania (40° North Latitude), or at Wellington, New Zealand (40° South Latitude), it is 800 miles per hour. At Bombay, India, or Mexico City, Mexico (20° North Latitude), or at Rio de Janeiro, Brazil (20° South Latitude), it is 975 miles per hour. Finally, at Singapore, Malaya, or Quito, Ecuador (on the equator), it is 1040 miles per hour.

The increasing centrifugal effect in a rotating planetary body lifts the very structure of the planet upward against gravity, the upward lift being greatest at the equator. In this manner, an "equatorial bulge" is formed, and the planet, instead of being a sphere, is an "oblate spheroid." For an oblate spheroid, the equatorial radius (the distance from the center of the planet to a point on the equator) is greater than the polar radius (the distance from the center of the planet to one of the poles). Naturally, as a given body is caused to rotate more and more rapidly, its equatorial bulge grows so that it departs more and more from a true sphere and becomes more and more oblate.

The oblateness of a planet is defined as the difference between the equatorial radius and the polar radius divided by the equatorial radius. For instance, Earth's equatorial radius is 3963 miles, while its polar radius is 3950 miles. The difference is 13 miles, and if that is divided by 3963, we find Earth's oblateness to be 0.0033, or about 1/300.

The actual oblateness depends not only on the velocity of rotation, but also, as we said above, on the density of the planet and the distribution of mass within it. In fact, it is possible to calculate what the oblateness ought to be for planets rotating at a particular velocity according to two different sets of assumptions: one, assuming the planet to be of uniform density throughout; and, two, assuming all the mass of the planet to be concentrated at the center. The first assumption (uniform density) yields markedly higher values for oblateness than does the second (concentrated mass). For instance, if the Earth were of the same density throughout, its oblateness should be about 1/250; if its mass were all concentrated at the center, it would be only about 1/500. The fact that the Earth's oblateness is intermediate between these two extremes indicates (as we know) that the Earth is not uniformly dense and that its mass is concentrated toward the center, but that the mass is not *all* at the center. The other oblate bodies of the Solar system also show values intermediate between the theoretical extremes, indicating that limited concentration is true for them as well.

The most oblate planets of the Solar system are Jupiter and Saturn. Despite their large size, their periods of rotation are considerably shorter than that of the Earth. Jupiter rotates in 9 hours, 50 minutes; Saturn rotates in 10 hours, 15 minutes. While a point on the Earth's equator moves at a velocity of 1040 miles per hour, one on Jupiter's equator (at the level of the cloud layer, that is, since that is all the surface we see) moves at 28,500 miles per hour, and a point on Saturn's equator moves at 23,000 miles per hour. The equatorial regions on those giant planets are lifted much higher than are those of the Earth, even against the stronger gravitational fields on Jupiter and Saturn. Jupiter's equatorial radius is nearly 3000 miles longer than its polar radius, while in the case of Saturn the difference is nearly 4000 miles. The oblateness of Jupiter is 0.062 or about 1/16, while that of Saturn is 0.096, or about 1/10. Earth's oblateness is so small that, seen from space, its shape would not depart visibly from a sphere. Jupiter and Saturn, as seen in a telescope, however, are clearly non-spherical, bulging amidships quite visibly.

The oblateness values of Uranus and Neptune are not known with precision. The oblateness of Uranus lies between 0.05 and 0.07 (1/20 to 1/14) while that of Neptune lies between 0.02 and 0.033 (1/50 to 1/20). Both planets revolve more rapidly than the Earth does, despite the fact that they are larger than the Earth, and both are, not surprisingly, considerably more oblate than the Earth.

Mars has a period of rotation about the same as that of the Earth, for it rotates in 24 hours, 37 minutes. Mars is a considerably smaller body and a point on its equator moves only 550 miles an hour, so it is not surprising that Mars has an oblateness too small to be measured precisely. The oblateness has been calculated, however, from slight changes in the orbits of Mars' tiny satellites and seems to be 0.0052 (about 1/200). This oblateness is greater than that of the Earth, despite the slower velocity of Mars' equatorial region, because Mars has a lower density than the Earth and is less compressed toward its center. Mercury, Venus, and the Moon, all with very slow periods of rotation, have no measurable oblateness.

Naturally, the surface gravity on an oblate spheroid would vary with position on the planet. A man standing on the Earth's equator would be farther from the center of the Earth than one standing at the North Pole, and the force of gravity would be consequently weaker at the equator. The effect is not marked. A man weighing 200 pounds at the North Pole (or the South Pole) would weigh only 199 pounds at the equator, if all measurements were corrected to sea level. The effect would be much greater, however, on planets as oblate as Jupiter or Saturn. On Saturn, gravitational attraction at the equator (at cloud level which, we again point out, is the only surface we can see), would be only four-fifths what it was at the poles. (A 200-pound weight at the Saturnian pole would weigh 160 pounds at the Saturnian equator.)

For the planets of our Solar system, a very interesting relationship between mass and rotation rate strongly suggests that these factors are not independent of each other. In general, the greater the mass, the more rapid the period of rotation. Thus, Jupiter, with the largest mass of any of the planets, also has the shortest period of rotation: 9 hours, 50 minutes. Saturn with a smaller mass rotates in 10 hours, 15 seconds. Uranus and Neptune, with masses still smaller, rotate in 11 or 12 hours. (Neptune's period may be as long as 15 hours.) Finally, Mars, which is far smaller than any of the outer planets, rotates in 24 hours, 37 minutes.

This relationship of mass and rotation period can be made a rather simple one by the proper mathematical treatment so that the rotation period of a body of given mass can be easily predicted. The Earth itself, as it turns out, has a longer rotational period than one would expect from this relationship. After all, it is 10 times as massive as

Mars, yet it rotates in about the same time. If the relationship displayed by the data obtained from the other planets held for the Earth, it would be rotating in 15.5 hours rather than in 24. However, it seems quite clear that over the ages, the Earth's rotation period has been slowed by the tidal effects of the Moon. The other planets, Mars, Jupiter, Saturn, Uranus, and Neptune, do not have any satellite as large in relation to themselves as the Moon is in relation to the Earth. They therefore have not suffered a comparable slowing effect.

The Moon itself suffers a slowing effect even greater than that suffered by the Earth. While the Earth is affected by the Moon's gravity, the Moon is affected by the Earth's 81-fold-greater gravitational field. The Moon's rotation has been slowed to a complete standstill with respect to the Earth, so that one of its sides always faces us. Its rotation with respect to the Sun, however, has not stopped. Its solar day is about 700 hours, which is equal to its period of revolution about the Earth (29½ days, or one synodic month).

The period of rotation of Mercury has been drastically slowed by the Sun's tidal effects and is now equal to 88 days, a period equal to its time of revolution about the Sun. The period of rotation of Venus is not known because of its featureless and obscuring cloud cover. In 1962, data obtained by reflecting radio signals from the surface of Venus were interpreted as indicating that Venus' period of rotation equaled its period of revolution (225 days). The indications are not conclusive, however, although we may be quite sure that Venus does rotate quite slowly as a result of the Sun's tidal pull. Were it not for the slowing effects of external bodies, Venus might be rotating in a period of 16 hours, Mercury in a period of some 30 hours, and the Moon in one of some 40 hours.

Spacing of Planets in the Solar System

The regular spacing of the orbits of the planets of the Solar system has interested students of astronomy ever since the planetary orbits were first established and measured with accuracy.

Prior to the mid-19th century, much significance was attached to the Titius-Bode law of orbital distances. This rule was expressed by a particular series of numbers, derived as follows: Starting with the number 3, follow it by additional numbers each twice the preceding number. Place a zero at the beginning of the list, and the result is 0, 3, 6, 12, 24, 48, 96, 192, 384, 768. . . . To each number in the series, starting with the zero, 4 is added. The new series thus produced is 4, 7, 10, 16, 28, 52, 100, 196, 388, 772. . . .

If one sets Earth's distance from the Sun equal to 10, the average distances of some of the other planets from the Sun (including that of Ceres, the largest of the asteroids), in proportion, are as follows:

Mercury	------------------	3.9
Venus	------------------	7.2
Earth	------------------	10.0
Mars	------------------	15.2
Ceres	------------------	28
Jupiter	------------------	52
Saturn	------------------	95
Uranus	------------------	192

In the 1840's, these were the only planets known, and the manner in which their relative distances fit the Titius-Bode law was very impressive. Neptune, however, is at a distance of 301 and Pluto at an average distance of 395, where the rule predicts 388 and 772, respectively. With the discovery of Neptune in 1846, therefore, the rule ceased to be regarded as anything but an interesting coincidence.

Another way of describing the planetary spacing is to make use of theoretical considerations of orbital stability and of the existence of forbidden and permitted regions. In general, the concept may be summarized as follows: Each planet, as it moves in its orbit, creates a broad band centered on that orbit (a "forbidden region") within which no smaller planet can exist in a stable orbit. If a smaller planet somehow found itself in the forbidden region, the gravitational pulls of the nearby larger planet would in a short time (astronomically speaking) move it out of the region.

The size of this forbidden region depends, in

part, on the distance of the planet from the Sun, so that the forbidden regions of successive planets tend to increase in width as one moves outward from the Sun. Thus, the width of the forbidden region associated with Jupiter is 250 million miles, whereas that associated with Saturn is 350 million miles. You would not expect the forbidden regions of adjacent planets to overlap, and Jupiter and Saturn are indeed spaced far enough apart to prevent this. In general, then, as one recedes from the Sun, the planets are spaced farther and farther apart, as the forbidden regions widen. The increasing numbers of the Titius-Bode law reflect this, and that is why the law seemed to work at first.

The width of the forbidden region also increases with mass, so that Jupiter's forbidden region is larger than it might be, judging only by its distance from the Sun. Uranus and Neptune, being distinctly smaller than Jupiter and Saturn, have forbidden regions that are correspondingly smaller than they might be, and this may account for Neptune's being closer to the Sun (and hence to Uranus) than would be predicted by the Titius-Bode law.

The width of the forbidden region also increases with the eccentricity of a planet's orbit; that is, the manner in which the shape of that orbit deviates from a perfect circle. For most planets, this deviation is quite small and can be disregarded. Mercury's orbit, however, and even more so, Pluto's, are distinctly elliptical. For that reason, both Mercury and Pluto have forbidden regions that are much wider than one would expect from their mass and their distance from the Sun.

The spacing of the planets about the Sun can then be viewed as resulting from the tendency of each planet to stake out a territory for itself. The first eight planets, Mercury, Venus, the Earth, Mars, Jupiter, Saturn, Uranus, and Neptune, are all situated so that no forbidden regions overlap. Pluto does not follow this rule. Although Pluto's average distance lies outside Neptune's forbidden region, Pluto's orbit crosses into it at Pluto's nearest approach to the Sun. (Since Pluto's orbit is markedly elliptical, it approaches far closer to the Sun at one region of the orbit than it would if its orbit were at the same average distance, but nearly circular.) Pluto's orbit is clearly unstable for this reason, and the situation may even foreshadow an eventual catastrophic alteration of Pluto's orbit by Neptune sometime in the distant future. Indeed, it may be that Pluto's existence as a planet is the result of such a catastrophe in the distant past. Some astronomers suspect Pluto to have been a satellite of Neptune at one time. In some fashion, it came to be thrown clear, but it has retained a highly unusual orbit, invading its ex-companion's forbidden region ever since.

The gaps between the forbidden regions are not always as narrow as they might be, so the forbidden regions take up only about 50 per cent of all the planetary space of the Solar system. Notably there is a wide gap between the forbidden regions of Mars and Jupiter, and within this gap, the asteroids circle. There are also interestingly wide gaps between the forbidden regions of Uranus and Saturn and between those of Neptune and Uranus, where small orbiting objects (as yet undiscovered) may well exist.

This pattern of regularity of the Solar system should also be found in other planetary systems.

Planetary

Habitability

In Chapter 2, we dealt with the range of environmental conditions that is consistent with habitability. In Chapter 3, we discussed the nature of the properties of the planets we know, those of our Solar system. Now it is time to see how these planetary properties relate to habitability, or lack of it.

Mass

The most basic of planetary properties is mass, since this determines, in great part (see page 55), a variety of other planetary properties, such as the density of the planet, its volume and radius, its surface gravity and escape velocity, its atmospheric composition and surface pressure, whether or not water is retained, the level of radio-activity at the surface, the topographical rough-ness, the rate of vulcanism, and so on.

In setting a maximum mass for a habitable planet, the limiting property would seem to be surface gravity. It will be recalled that to be con-sidered habitable, a planet must have a surface gravity of not more than 1.5 g (see page 24). A surface gravity of this magnitude is produced by a planet that has a mass 2.35 times that of the Earth. Such a planet would have a radius 1.25 times that of the Earth. It would, in other words, have a diameter of 10,000 miles and a circumference of 31,400 miles, as compared with corresponding fig-ures of 8000 miles and 25,000 miles for the Earth.

What about the lower limit on mass? Here we cannot use surface gravity as a direct criterion, for there is no known lower limit on the intensity of the gravitational field as far as human physi-ology is concerned. Instead, we will ask ourselves at what point the mass of a planet becomes too small to retain a breathable atmosphere on its surface.

As we have seen (see page 27), the mini-mum tolerable atmospheric pressure for human beings is some 2 psi, provided the atmosphere is pure oxygen. Some nitrogen, however, is necessary for plants. Hence, if we assume an atmosphere consisting of, say, 90 per cent oxygen and 10 per cent nitrogen (plus a bit of water vapor and a trace of carbon dioxide), the minimal barometric surface pressure required of an atmosphere would be about 2.3 psi, or 0.156 atmosphere.

If we assume that all habitable planets must have surface temperatures in the approximate neighborhood of those on the Earth, we can conclude that their exosphere temperatures will also be similar to that of the Earth. We can calculate the rms velocity of oxygen atoms at such an exosphere temperature. (Oxygen, as it occurs in the body of the atmosphere, is in the form of molecules, each of which is made up of a pair of atoms. In the exosphere, however, photolysis breaks up the molecule into individual atoms. The oxygen atom, being only half as massive as the two-atom molecule, has a higher rms velocity than the molecule. Even if the planet could retain the molecule, it might still be unable to retain the individual atom and would lose its atmosphere by way of those atoms.)

In order to retain the oxygen atom, the escape velocity of the planet must be, as a bare minimum, at least 5 times as high as the rms velocity of the oxygen atom (6 times as high would be safer). This means that the escape velocity of the smallest planet capable of retaining atomic oxygen must be no lower than 4 miles per second. Such an escape velocity will be found on a planet having a mass 0.195 times that of the Earth, a radius 0.63 times that of the Earth, and a surface gravity of 0.49 g.

This, however, is not satisfactory. Conclusions based on escape velocity tell us only that such a planet could hold an oxygen-rich atmosphere if it had one to begin with. The question is, however, whether a planet having only one-fifth the mass of the Earth would have formed an atmosphere of this sort in the first place. Remember that it is not only an atmosphere that we require (vulcanism would take care of that), but an atmosphere high in free oxygen (requiring more than vulcanism).

There are several processes that must take place before a planet can have an oxygen-rich atmosphere of a density consistent with habitability. First, there must be some mechanism for the production of free oxygen. Second, there must be some mechanism for the accumulation of free oxygen in the atmosphere to an inspired partial pressure of at least 60 mm. Hg.

On the Earth, the existence of free oxygen in the atmosphere can probably be attributed entirely to the photosynthesizing activity of green plants, which use light energy to split the water molecule into hydrogen and oxygen. The hydrogen is combined with carbon dioxide to form carbohydrates and plant tissue generally, while the oxygen is liberated into the atmosphere.

However, there are a number of processes tending to remove free oxygen from the atmosphere after it is formed. Oxygen is consumed by the oxidation of minerals in the crust during weathering, by the respiration of animals, and by the oxidation of plants when they die and decay. When a plant decays or is burned completely, it uses up just as much oxygen as it produced while it was growing. While the plants are alive, their mass is balanced by an equivalent mass of free oxygen in the atmosphere. It is estimated, however, that the mass of plant life on the Earth is only about one-sixtieth the mass of the oxygen in the atmosphere. Some explanation must be sought for the accumulation of oxygen to an extent 60 times the mass of plant life in the face of losses to weathering and animal respiration.

What would seem to be required would be the progressive withdrawal of plant life from the Earth's surface and its permanent protection against the possibility of oxidation. The oxygen-contribution of such plants is then preserved indefinitely in the atmosphere. As a matter of fact, vast quantities of organic matter have, during the history of the planet, been buried or submerged in sediments and appear now, unoxidized, as coal and oil. This is apparently the major process through which free oxygen has been accumulated in the Earth's atmosphere.

Intuitively, one would expect that small planets would have a lower rate of burial of organic matter than Earth has. Furthermore, a large share of the photosynthesis taking place on Earth is carried out in the oceans (over 70 per cent, according to some estimates). One would therefore expect that a low sea-land ratio would lower the amount of photosynthesis and therefore the proportion of free oxygen in the atmosphere. (To be sure, there might be a gain in land area, but with the shrinkage of ocean, most of that land area would be desert and would not contribute at all to photosynthesis.) Smaller planets, with a lessened capacity to produce and retain water, would be expected to have relatively low sea-land ratios. From these considerations, it can be suspected that a planet just large enough to retain a breathable atmosphere may nevertheless be too small to form one in the first place. We will therefore reject a mass 0.195 times that of Earth (as obtained from escape-velocity considerations) as representing too low a minimum.

Let us begin again, then, by taking two known cases, those of the Earth and Mars. Earth's atmosphere has a surface barometric pressure of 14.7 psi, while that of Mars' atmosphere is estimated to be, at most, one-seventh of this figure, or 2 psi. Earth's atmosphere is 21 per cent oxygen, while that of Mars has virtually no oxygen. The mass of Mars, finally, is 0.11 times that of Earth. We can imagine that as the mass of Mars is increased, its atmosphere increases, too, both in total surface pressure and in oxygen content. We can ask ourselves, then, what mass a planet like Mars must attain before the barometric pressure of its atmosphere becomes 8 psi and the oxygen content 0.16, these values yielding the necessary minimum inspired oxygen partial pressure of 60 mm. Hg.

We can calculate this minimum mass on the

basis of two possible assumptions: (1) that atmospheric pressure will be proportional to the planet's surface gravity and (2) that atmospheric pressure will be proportional to the planet's mass. On the basis of the first assumption, the minimum mass turns out to be 0.25 times that of the Earth; and on the basis of the second assumption, the minimum mass appears to be 0.57 times that of the Earth.

Since it is unlikely that either extreme is entirely true, we shall make the reasonable supposition that the value we really want is somewhere between 0.25 and 0.57. Let us say, then, that the minimum mass of a habitable planet is 0.4 times that of the Earth. This corresponds to a planet having a radius 0.78 times that of the Earth and a surface gravity of 0.68 g. Such a planet would have a diameter of 6200 miles and a circumference of 19,400 miles.

To summarize, then, we will postulate that the mass of habitable planets may vary over the range of 0.4 to 2.35 times that of the Earth; the radius may vary from 0.78 to 1.25 times that of the Earth; the diameter from 6200 miles to 10,000 miles; the circumference from 19,400 miles to 31,-400 miles; and the surface gravity from 0.68 to 1.5 g.

Of the planets of the Solar system, only Earth and Venus fall in this range. Mars, Mercury, and the various satellites and asteroids all fall well below the minimum; while Jupiter, Saturn, Uranus,

	Smallest	The Earth	Largest	Moon	Mars
Mass	0.40	1.00	2.35	0.0123	0.1077
Radius	0.78 / 3090 mi	1.00 / 3960 mi	1.25 / 4950 mi	0.273 / 1080 mi	0.53 / 2100 mi
Surface gravity	0.68	1.00	1.50	0.165	0.38

and Neptune are all well above the maximum. (The situation in regard to Pluto is uncertain because its mass is not known with any degree of precision.)

Although the permitted mass range for habitability may seem narrow compared to the mass range actually observed for planets, there is still room for considerable variation in properties within the habitable mass range. Generally speaking, planets near the lower end of the permissible mass range might be expected to have developed lower internal temperatures during their period of formation, to have cooled more rapidly, and to have thicker crusts. They would exhibit less stratification within the crust; that is, there would be a smaller concentration of dense materials at their centers and of light materials in the crust (since the tendency for such concentration lessens as the gravitational field grows weaker). Consequently, the lighter planets may have crusts that are relatively richer in certain heavy metals.

Lighter planets might also be expected to show less volcanic activity and, consequently, to have smaller oceans and atmospheres. The smaller planet would have a "softer" atmosphere (see page 74), and the density of its atmosphere would fall off less rapidly with height. One effect of this is that at high altitudes (above, say, about 30,000 to 40,000 feet) the smaller planets should have denser atmospheres than the larger planets. This would have a distinct influence on such factors as the altitude ceilings of similar kinds of aircraft flying in the atmospheres of planets with different masses.

Rate of Rotation

The rate of rotation of a planet is a key factor in determining its oblateness (see page 76). Other characteristics also depend in part on rotation rate—such as change of surface gravity with latitude, the daily temperature cycles, atmospheric circulation patterns and wind velocities, and possibly the magnetic field.

In general, the slower the rotation rate, the greater the day-to-night temperature differences would be. From the standpoint of habitability, a lower limit on the rate of rotation would be reached when daytime temperatures became excessively high in the regions below some definite latitude and nighttime temperatures became excessively low above that same latitude. A lower limit might be reached before this point when the light-darkness cycle became too slow to enable plants to live through the long hot days and long cold nights.

At the opposite extreme, that of rapid rotation, a limiting point would be reached when centrifugal effects caused surface gravity at the equator to fall to zero so that matter would be lost from the planet, or when the shape of the surface became unstable and symmetry about the planet's axis was lost.

Just what extremes of rotation rate actually represent these limits of habitability is difficult to say. They might be estimated, however, at 96 hours (4 Earth days) per rotation for the lower end of the scale and 2 to 3 hours per rotation for the upper end.

A special case that might be considered is one in which the planet's period of rotation is precisely equal to its period of revolution about its sun, so that one side of the planet is in perpetual light while the other side is in perpetual darkness. This might be thought to be compatible with habitability over at least a limited region of the planet, since temperatures near the day-night line (or "twilight zone") might be in the desired range.

However, just what would happen to a planet's atmosphere under these circumstances? Would the atmospheric circulation be strong enough to prevent all the gases from condensing on the dark side? Or would all the water and carbon dioxide, at least, precipitate out in the extreme cold of the dark side? If it is assumed, as seems reasonable, that the day-equals-year situation was not present from the very start but was preceded by a long slowing-down period, then during that period all the planet's water might well have been converted by photolysis to hydrogen and oxygen during the increasingly long and hot day, with the subsequent escape of hydrogen. The day-equals-year case may be ruled out, then, as indicating either that all the water is precipitated out in solid form on the dark side or that the planet is completely dry. In any case, there would be no open stretches of liquid water on such a planet, not even in the "twilight zone," and consequently, it would not be habitable according to the definition of the term used in this book.

Considering high rotation rates again, it is apparent that the force of gravity varies with latitude, being lowest at the equator and highest at the poles (see page 80). For an Earth-like planet with a 3-hour period of rotation, for example (if it is assumed that the average density of Earth is unchanged), the oblateness would be about 0.24 (or 1/4), and the force of gravity at the equator would be about 0.7 g.

It might be thought that this effect could extend the upper limit of mass of a habitable planet. A planet with too great a surface gravity, on the average, may yet be rotating so unusually rapidly that the surface gravity may fall to habitable levels in its tropical regions. This may indeed have some marginal effect; and a planet with a surface gravity somewhat higher than 1.5 g, on the average, might conceivably be rotating rapidly enough to bring it down to 1.5 g or below in the tropics.

This, however, would be a very rare situation. We may speak of rapid rotation rates of 2 to 3 hours for planets in the habitable mass range, but how likely is it for such rapid rates actually to be attained? Based on data on the planets of our Solar system, it seems quite probable that rate of rotation varies roughly with mass (see page 81). Planets in the habitable range of mass would, in all likelihood, have rotation periods of 15 to 20 hours, and this is insufficient to introduce any marked difference in surface gravity with latitude. This period can be slowed easily enough by the tidal influence of a satellite as, in actual fact, Earth's period of rotation has been. The chance of a considerably more rapid rotation rate, however, would be very slight indeed.

For a rotation rate to be large enough to introduce a marked lowering of gravitational attraction in the tropic regions (as in the case of Saturn, for instance), the planet must be so massive as to be far outside the habitable range; it must be so massive, in fact, that the lowering of gravity with latitude is not likely to reduce it to habitable levels. To be sure, the surface gravities of Saturn, Uranus, and Neptune are usually given as between 1 and 1.5 g; but this is not because of the effect of rapid rotation. It is because the "surface" referred to is actually the top of a cloud layer that may be hundreds or even thousands of miles above the solid surface. The surface gravity on the solid planet itself is unknown, since we do not know how far that solid surface is from the center of the planet. It seems certain, however, that the gravity at the solid surface would be well over 1.5 g.

Age

A certain amount of time must elapse before a newly formed planet can have surface conditions suitable for life. The sequence of events for an Earth-like planet, over the course of its history, might be something like this:

1. A planet is formed by the gradual accretion and capture of small particles.

2. After the accretion process has ended because of a lack of growth materials, the surface is airless, or very nearly so.

3. The interior of the planet is extremely hot as a result of gravitational compression. Internal readjustments are taking place: denser materials, such as iron, are flowing slowly downward, and lighter materials, such as the various rocky silicates, are flowing slowly upward. Because of the high viscosities of the materials involved, the internal readjustments take place over a long period of time. They also produce movements in surface materials, with extensive vulcanism, crustal movements, and earthquakes. Localized heating of crustal rocks by friction and through the decay of radioactive materials in the mantle produces high temperatures, and trapped gases are released.

4. The lighter gases (hydrogen, helium, neon)

escape from the planet altogether after they are released, but the heavier gases are retained. Gases such as methane and ammonia may begin to accumulate if their rate of evolution exceeds their rate of escape. Much of the water vapor that is produced is broken down by photolysis into hydrogen and oxygen. The hydrogen escapes from the planet, and the oxygen enters into chemical reactions with surface materials.

5. Those stable gases that can be retained (nitrogen, carbon dioxide, and, possibly, methane and ammonia) begin to accumulate, and an atmosphere starts to build up.

6. As the atmosphere becomes thicker, and volcanic action continues, a point is reached when the rate of production of water vapor exceeds the rate of loss by photolysis. Furthermore, as oxygen continues to be produced by the photolysis of water, and the surface material is largely oxidized, traces of free oxygen (and its more active form, ozone) can begin to accumulate in the atmosphere.

7. The oxygen and, particularly, the ozone absorb light in the ultraviolet region of the solar spectrum (those wavelengths that happen to be responsible for the photolysis of water). As a result, photolysis slows down and water vapor can now begin to accumulate more rapidly. The presence of even small amounts of ozone also produces a stable upper level of the atmosphere so that water vapor is unable to diffuse upward so rapidly. This is known as the atmospheric "cold trap," and

results in the freezing and precipitation of water. Since water vapor cannot escape from the planet unless it makes its way into the upper atmosphere first, this is an important factor in the retention of water vapor.

8. When the concentration of water vapor in the atmosphere reaches the dew point or frost point, liquid water or solid frost condenses out locally. When the atmospheric pressure has become high enough (and assuming that the temperature remains above the freezing point of water), water begins to accumulate on the planet's surface.

9. With the beginnings of ocean formation and with continuing vulcanism, most of the carbon dioxide goes into solution, forming carbonic acid and reacting to form carbonate rocks. The ammonia also goes into solution and enters into reactions. Now the atmosphere consists mainly of nitrogen and methane, plus small quantities of carbon dioxide, and with water vapor as a variable constituent.

10. The oceans increase in size so that rainfall becomes more prevalent. Weathering begins to become significant and soluble minerals are washed into the oceans.

11. More complicated chemical species begin to accumulate in the oceans, the process being fed by energy absorbed from sunlight. Lightning discharges form small quantities of the oxides of nitrogen; these dissolve to form nitric acid and

nitrates. Sulfur dioxide from volcanoes dissolves to form sulfuric acid and sulfates.

12. At some point, life appears, and, eventually, photosynthesis is established. Through photosynthesis, oxygen begins to accumulate in the atmosphere at a rate far greater than would be possible through photolysis alone.

13. After a long period of time during which the prevalence of photosynthesizing organisms increases, the oxygen concentration of the atmosphere reaches the minimum value required by land vertebrates. The volcanic activity level slows down, the meteorite-infall rate diminishes, and the planet may be considered habitable.

How long does this entire process take? A billion years? Two billion? Three billion? It is not possible to say with much accuracy, but the amount of time is surely of this order of magnitude. Thus, even though a planet has all the other essential attributes from an astronomical point of view, it must also be of a certain age; it must have ripened sufficiently, so to speak, before it can be considered habitable.

From the evolutionary point of view (as shown in the sketchy chronological sequence given above), it may be seen that several factors could interfere with the development of suitable conditions on the surface of a planet. If the planetary mass were somewhat too small, the rate of water production by volcanic activity would be too low to balance the rate of loss by photolysis, and water might never accumulate on the surface. (This seems to have happened to Mars, for instance.) If the mean surface temperature were too high, water would never condense on the surface; instead, all of it would remain in the atmosphere where it would continue to be lost slowly by photolysis. No oceans would form, and carbon dioxide would become a major constituent of the atmosphere. (This seems to have happened to Venus.)

In general, it is probably safe to say that a planet must have existed for 2 or 3 billion years, under fairly steady conditions of solar radiation, before it has matured enough to be habitable. It may be estimated very roughly that the Earth itself is 4.5 to 5 billion years old and that traces of life upon it extend as far back, perhaps, as 2.5 to 3 billion years. From this it would follow that the Earth was some 2 billion years old before life developed on it. If we assume that life would increase in complexity and spread out over the planet enough to increase the oxygen content of the atmosphere over the next billion years, we can say that Earth had to be 3 billion years old before it could be considered potentially habitable. We will therefore in this book accept 3 billion years as the minimum age of a habitable planet.

Distance from Primary and Inclination of Equator

In astronomical parlance, a "primary" is a body about which a second and less massive body revolves. The Earth is the Moon's primary, and the Sun is the Earth's primary. The most massive star of a multiple system is the primary of that system. In this book, however, we will use the word exclusively for the star that supplies the major portion of the heat and light for the non-luminous bodies making up the rest of the stellar system. From this viewpoint, the Sun is the Moon's primary as well as the Earth's.

The word "star" can be used as a synonym for "primary," and at times it will be; but ordinarily the picture conjured up by "star" is that of a small bright dot in the night sky, rather than of a huge, radiant sun. The word "sun" can also be used as a synonym for "primary," and at times it will be; but ordinarily the picture conjured up by "sun" is that of our own Sun in particular. In most instances, then, we will use the word "primary."

In turning now to the business at hand, we are dealing with two factors. The first is the distance of a planet from its primary. The second is the inclination of the plane of its equator to the plane of its orbit of revolution around the primary. (This latter can also be pictured as the "tipping of the axis.") These factors must be considered together because habitability depends on the two in combination rather than on each independently. Orbital eccentricity (to be discussed shortly) is also interrelated with these factors in determining habitability, but at the moment, let us assume that

the orbital eccentricity is zero; that is, that the
orbit of the planet is a perfect circle centered on
a star.

The radiation received by a planet from a
particular star (the planet's illuminance, in other
words) depends on the distance of the planet from
that star; and the habitability or non-habitability
of the planet depends on the quantity of illuminance. The question of the planet's distance from
the primary is therefore crucial.

To be exact, the illuminance varies with the
square of the distance of a planet from its primary.
If one planet is twice as far from the primary as
a second planet, it receives 2^2, or 4, times less
illuminance. If it is 5 times as far, it receives 5^2, or
25, times less illuminance. (Earth's illuminance,
incidentally, measured at the top of its atmosphere
is 1.94 gram calories per square centimeter per
minute. This is usually called the "solar constant.")

In connection with illuminance, it will be useful to introduce the term "ecosphere." For present
purposes, ecosphere will be used to mean a region
in space, in the vicinity of a star, in which suitable
planets can have surface conditions compatible
with the origin, evolution to complex forms, and
continuous existence of land life; and, in particular, surface conditions suitable for human beings,
together with the whole system of life forms on
which they depend.

The ecosphere is bounded by two spherical
shells, centered on the primary. Inside the inner

shell, illuminance levels are too high for habitability; outside the outer shell, they are too low. In general, then, we can say that to be habitable, a planet must be inside the ecosphere.

It is a difficult problem to predict temperatures at a particular location on the surface of a planet as functions of illuminance and of equatorial inclination. The problem becomes extremely complicated when a planet has atmospheric circulation and irregularly shaped ocean and dry-land areas, as a habitable planet would be expected to have. Even the attempts made to calculate the mean annual temperatures on the Earth's surface (concerning which so much is known) on purely theoretical grounds have not been highly successful. How then can we apply theory to habitable planets whose surfaces we cannot know in detail? Because of the difficulties involved, we have had to use empirical methods for determining planetary surface temperatures, using the Earth as a standard.

We assumed, first, that we were dealing with Earth-like planets having thin, transparent atmospheres and a cloud cover of approximately 45 per cent. Theoretical temperatures were then calculated at various latitudes and seasons for rapidly rotating, non-conducting black spheres of various equatorial inclinations that were half illuminated by a distant point source of light. (This is a highly idealized and simplified version of the Earth-Sun relationship.) Once a list of theoretical temperatures was obtained, this was compared with the actual observed temperatures on the Earth's surface, and the theory was modified to fit. In this manner, average surface temperatures were estimated for planets at different illuminances and different equatorial inclinations. These were estimated for various latitudes at various times of the year: summer solstice, winter solstice, and the equinoxes.

Finally, habitability was judged by applying the rule (see page 16) that a region is habitable only if the mean annual temperature lies between 32° F. (0° C.) and 86° F. (30° C.), if the highest average daily temperature in the hottest season is less than 104° F. (40° C.), and the lowest average daily temperature is higher than 14° F. (−10° C.).

From such calculations, it turns out that for a planet to have at least 10 per cent of its surface fall within the habitable temperature range, equatorial inclinations up to approximately 80° are tolerable.

In this connection, it might be pointed out that at an equatorial inclination of 0°, the planet's axis is perpendicular to the plane of its revolution about the primary, and every part of the planet has days and nights of equal length throughout the year. Where the axis is "tipped," the lengths of days and nights vary with the time of year (as in Earth's case, where the equatorial inclination is 23.5°). This is the cause of seasons.

At an equatorial inclination of 90°, the planet's axis lies in the plane of revolution. Whatever the rate of rotation about the axis, the surface of the planet receives a complicated annual pattern of sunlight. When one pole is pointed at the primary, that hemisphere receives sunshine all day long while the other hemisphere is in darkness. A quarter of a year later, the axis is broadside to the primary, and all parts of the surface have equal days and nights. After another quarter year, the other pole is pointing at the primary; thus each pole is alternately baked and frozen, while the equatorial belt has two frigid seasons every year with the sun near the horizon for days on end, and two more "normal" warmer seasons in between. Equatorial inclinations of more than 90° duplicate the conditions for inclinations of correspondingly less than 90° except that the planet now rotates from east to west rather than from west to east. Climatically speaking, an equatorial inclination of 100° is equivalent to one of 80°; one of 135° is equivalent to one of 45°; and one of 180° is equivalent to one of 0°.

By our calculations, then, it is only in the narrow range of equatorial inclination between 80° and 100° that no position within the ecosphere can be found that will meet the temperature requirements for at least 10 per cent of the planet's surface. For that range of equatorial inclination, then, a planet must be considered non-habitable, however well it meets all other requirements. (The planet Uranus, with an equatorial inclination of 98°, is the only body known to fall into this range. Uranus, however, is non-habitable for many other reasons as well.)

Total illuminance can be consistent with habitability over the range from 0.65 to 1.9 times that received at the Earth. At every point throughout this range, a planet can exist at some appropriate equatorial inclination that will place at least 10 per cent of its surface within the required temperature range consistent with habitability. (For moderate equatorial inclinations, up to 54°, a maximum illuminance of 1.35 times Earth normal is the limit.)

These figures enable a boundary to be placed on the ecosphere of any particular star. (Naturally, a bright, hot star will have an ecosphere much farther out in space than will a dim, cool star.) In our own Solar system, the ecosphere extends from 0.725 A.U. (where the illuminance is 1.9 times that of the Earth) to 1.24 A.U. (where it is 0.65 times that of the Earth). In miles, with 1 A.U. equaling 93 million miles (see page 4), the ecosphere extends from 67.5 million to 115 million miles from the Sun. The inner edge of the ecosphere reaches the orbit of Venus (which is at a distance of 0.723 A.U. from the Sun), and its outer boundary reaches halfway to the orbit of Mars (which is at an average distance of 1.526 A.U. from the Sun).

Of course, the determination of the boundary of the ecosphere has rested on so many assumptions that the figures given can only be considered as approximate.

Orbital Eccentricity

The orbits of the planets have the shapes of ellipses, these being closed curves that resemble flattened circles. The Sun is not at the center of such an elliptical orbit but is to one side of the center at a point called the "focus."

In a circle, all straight lines passing through the center from one point on the circumference to another ("diameters") are equal. In an ellipse, such lines are not equal. There is one that is shorter than any of the others and this is the "minor axis." At right angles to the minor axis is the longest diameter, and this is the "major axis." The two foci of an ellipse are located on the major axis, one focus to one side of the center, the other focus to the other side.

The more flattened the ellipse, the farther the foci are located from the center and the greater the difference in length between the major axis and the minor axis. The distance between one focus and the center, divided by the length of the major axis, is the "eccentricity" of the ellipse. The eccentricity can be as low as 0, in which case there is no difference in length between the axes so that the ellipse is completely unflattened and, therefore, is a circle. The eccentricity can be as high as 1, in which case the ellipse is either completely flattened into a straight line or else is lengthened infinitely into a curve called a "parabola."

The Sun, as stated above, is at one focus of an elliptical orbit (the other focus is empty), and the major axis of the orbit therefore passes through the Sun. When the planet passes the end of this major axis, it is either at a point closest to the Sun (if that end of the major axis is on the same side as is the focus occupied by the Sun) or at the point farthest from the Sun (if it is at the end of the major axis on the side opposite that of the focus occupied by the Sun). The more eccentric the orbit, the more flattened is the ellipse, the farther located to one side of the center is the Sun, and the greater is the difference between the point of closest approach to the Sun at one end of the major axis and the point of greatest distance at the other.

(The point of least distance to the Sun is the "perihelion," while that of greatest distance is the "aphelion." When a star is involved, the two terms are "periastron" and "apastron.")

The eccentricity of planetary orbits within the Solar system is generally low. That of the Earth's orbit is 0.017, so that at perihelion it is 91.3 million miles from the Sun, and at aphelion, 94.5 million miles. This difference of 3.2 million miles may seem great, but in comparison with the over-all size of the orbit it is not significant. If the Earth's orbit were drawn to scale it could not be told from a circle by the unaided eye. The orbits of Neptune and Venus are even less eccentric; they have eccentricities of 0.009 and 0.007, respectively. The only planetary orbits with eccentricities higher than 0.1 are those of Mercury (0.206) and of Pluto (0.247), the innermost and outermost planets, respectively.

Those satellites that we have included among the planets by the definition used in this book (see page 58) also have low orbital eccentricities. The Moon's orbit about the Earth has an eccentricity of 0.055, for example. (For objects other than planets, orbits of marked eccentricity are much more common. Many asteroids have orbits with eccentricities higher than 0.2, while many comets move in orbits with eccentricities higher than 0.9.)

The seasonal temperature cycle will vary by hemispheres if the planet happens to reach aphelion and perihelion at the time of the solstices. This state of affairs is actually to be found on the Earth: The Earth reaches aphelion, and is farthest from the Sun, around July 4, shortly after the beginning of winter in the Southern Hemisphere and summer in the Northern Hemisphere. Because of the unusual distance of the Sun, the Southern Hemisphere winter is colder than it might be, and the Northern Hemisphere summer is less hot than it might be. The Earth then reaches perihelion, and is nearest the Sun, around January 4, after the beginning of summer in the Southern Hemisphere and winter in the Northern Hemisphere. Because

of the unusual closeness of the Sun, the Southern
Hemisphere summer is hotter than it might be,
and the Northern Hemisphere winter is less cold
than it might be. In other words, the fact that
perihelion and aphelion fall quite close to the
solstices gives one hemisphere (the Southern, in
our case) extreme seasons, and the other (the
Northern), mild seasons. The effect in the Earth's
case is slight and subordinate to other factors, but
if the orbit were more eccentric the effect might
be considerable. The nature of the effect might
also depend on the period of the year. Planets
with a short year might remain habitable even
with relatively high values of eccentricity because
seasonal changes would tend to become blurred
by the natural sluggishness of response, or "lag,"
in seasonal change. Planets with long years would
be more seriously affected by high values of orbital eccentricity.

The association of the perihelion and aphelion
with particular seasons of the year does not remain fixed, however. The time of the solstice
varies slowly because of the shift of direction of
the planetary axis. This direction slowly describes
a circle with respect to the stars. It is this "precession of the equinoxes" that causes different
stars to serve in turn as "pole stars." Because of
this precession, there are, inevitably, times in the
history of a planet when aphelion and winter
solstice coincide for one hemisphere. Half a precession period later, the situation is reversed, and

the other hemisphere gets the extremes. The period of the precession of the Earth's axis is 25,600 years. This means that about 13,000 years hence, it will be the Northern Hemisphere that will be getting the seasonal extremes and the Southern that will be more moderate.

Some trial calculations, using the Earth as an example and assuming that aphelion coincided with winter solstice, indicated that habitability is not affected in any significant manner by eccentricities up to 0.2, and it seems reasonable to accept that value as an arbitrary upper limit for the orbital eccentricity of habitable planets. (Greater eccentricities appear, in any case, to be relatively improbable for bodies of planetary mass.)

Properties of the Primary

From the previous discussion of the dependence of habitability on planetary age (see page 106), it follows that a primary star must emit light and heat at a fairly constant rate for a period of at least 3 billion years. In order to go into this matter further, it will be useful now to give a very brief review of the classification of the stars.

Although all the stars, except our Sun, are so far away that they cannot be seen as disks but merely as points of light, a great deal of information has been accumulated about them from measurements of their luminosities, distances, and surface temperatures; from studies of the bright and dark lines in their spectra; and from measurements made with various instruments such as the interferometer, which enables astronomers to measure the diameters of certain very voluminous stars. All but a minor fraction of the stars (less than 1 per cent) belong to the main sequence (see page 42).

According to presently accepted views of stellar evolution, stars in the main sequence are in the stable phase of their existence and are converting hydrogen into helium at a steady rate. After they have consumed a certain fraction of their available hydrogen, stars leave the main sequence, expand greatly to become "red giants," and then go through various rapid evolutionary phases (sometimes including stages where their level of radiation changes rapidly and rhythmically so that they are "variable stars"). Stars eventually end as tiny "white dwarfs," small in volume but large in mass and tremendously high in density. The conversion to a white dwarf may be preceded by an explosive loss of mass, during which there is an overwhelming, though temporary, increase in brightness, and the star may go through a "nova" or "supernova" stage. The stars that are not on the main sequence are not steady enough in the radiation they deliver to serve as primaries for a habitable planet which, above all, needs a steady supply of radiation. In this book, therefore, we will concentrate on those stars (the vast majority, remember) that lie on the main sequence.

Nearly all stellar spectra can be arranged in a sequence marked by a smooth and continuous change in the intensities of the absorption lines, which reflect, we are certain, a smooth and continuous change in the surface temperature of the stars concerned. The spectral sequence contains seven main groups or classes designated (from hottest to coolest) as O, B, A, F, G, K, and M. The subdivisions of the groups are indicated by numbers from 0 to 9 following the letter—for example, B0, B1, B2, and so on. Our Sun is classified as G0 or, sometimes, as G2.

Among main sequence stars, the class O stars (very rare) are the most massive and the hottest, with surface temperatures up to 90,000° F. (50,000° C.). They have the largest diameters and the lowest densities. The class M stars (very abundant) are the least massive and the coolest, with surface temperatures as low as 5400° F. (3000° C.). They have the smallest diameters and the highest densities. The B, A, F, G, and K stars are intermediate with respect to these properties.

Stars use hydrogen according to their masses. If Star A is twice as massive as Star B, it has twice the quantity of hydrogen, but it consumes it at more than twice the rate. Consequently, the more massive a star, the more rapidly it uses up its hydrogen fuel, and the sooner it is forced to leave the main sequence. Stars of spectral class O spend very short times (astronomically speaking) on the main sequence, while stars of spectral class M spend very long times there. Again, the other classes are intermediate in this respect.

For a star to use its hydrogen at a slow enough rate to ensure a stay of at least 3 billion years on the main sequence (to give it time to develop habitable planets) it must have a mass not more than about 1.43 times that of our Sun. This means that to have habitable planets, a star cannot have a spectral class "earlier" than F2. Stars of spectral classes O, B, A, F0, and F1 cannot be expected to have habitable planets (though they may have planets). They just don't have the time for it.

It is also possible for a star to have too little mass to develop habitable planets. This is not (as one might think) because the insufficiently massive star is too cool to keep a planet warm enough for habitability. After all, the ecosphere might be imagined as drawn close about the star. The lower limit of mass is set, instead, by matters involving the braking of periods of rotation. A few words on this subject are necessary.

We have already explained that low rates of rotation are incompatible with habitability (see page 98) and that the effect of tides raised on one world by the gravitational influence of another is to slow the rotation rate. In the long run, the rotation rate is slowed to the point where it is stopped with respect to the other body. Thus the Earth's rotation has been slowed by the Moon's tides, and the Moon's rotation has been slowed even more by the Earth's tides. The Moon faces us with one side only. Similarly, Mercury and perhaps Venus face the Sun with one side only.

The study of the tides is not a simple one. On the Earth, tides in the middle of the ocean are only a foot or so in height above mean sea level, but the fact that the water in the oceans is confined in more or less rigid connecting basins with highly irregular rim and bottom shapes, means that water can pile up to heights of many feet on shore and in bays. Tides may be as high as 50 feet in the Nova Scotian Bay of Fundy. The details of water flow induced by tidal influence are very complicated indeed.

This flow of water against the bottoms and sides of the various shallow seas of the Earth dissipates rotational energy through friction, and this is sufficient to account for the observed slowing of the Earth's rotation (1 second every 100,000 years). Other factors may also be important, such as bodily tides in the solid structure of the Earth and changes in the Earth's moment of inertia due to shifts of matter within its solid structure. Also to be studied are changes in the oceans or in the sea level, tides in the atmosphere, and interactions between magnetic fields of Earth and Sun.

Naturally, most of this data is hard enough to obtain for the Earth and impossible, at present, to obtain for other planets. However, one theoretical analysis indicates that the rotation-retarding effects of tides is proportional to the square of the maximum height of the tides in deep water. We can symbolize this quantity as h^2.

If we take the value of h^2 on Earth due to the Moon to be 1, we can calculate what it might be on other planetary bodies. To be sure, h^2 has literal significance only for the Earth, since, as far as we know, the Earth is the only body in the Solar system that has open stretches of water to be affected by tides. Nevertheless, we can suppose that high values of h^2 will indicate strong tidal effects on atmospheres or on the solid structure of the planet that will have rotation-slowing effects of the same order of magnitude as ordinary tides would have if the planet had oceans.

With this thought in mind, we can see that most or all of the satellites of the Solar system, circling, as they do, in the vicinity of much larger bodies, must have their rotation stopped by tidal effects, at least with respect to the large bodies they circle. The value of h^2 produced by Jupiter on its nearest large satellite, Io, is 45,000,000; that produced by Neptune on its large satellite, Triton, is just about 500,000; and that of Saturn on its comparatively distant largest satellite, Titan, is 61,500. Surely each of these revolves about its planet in such a way as to keep the same side faced toward it at all times.

Of all the large satellites in the Solar system, the one that revolves about the smallest planet is our Moon. The Earth is far less capable of slowing its rotation than Jupiter, Saturn, or even Neptune would be. In fact, the value of h^2 produced by the Earth on the Moon is only 1325, but this is quite enough to have stopped the Moon's rotation with respect to the Earth.

As for the effect of the Sun on its planets, this braking effect decreases with distance. The value of h^2 produced by the Sun on Mercury is 9.65, and this is enough to ensure Mercury's keeping one face turned always to the Sun. For Venus, the value of h^2 is 1.77, and recent information obtained from the Venus probe, Mariner II, makes it seem that Venus, too, keeps one side to the Sun, or rotates very slowly at best. The value of h^2 produced on Earth by the Sun is 0.206, but that produced on the Earth by the Moon is 1.00. Consequently, the total value of h^2 on Earth is 1.2. This is enough to slow the rotation of the Earth but not enough to have stopped it so far.

The value of h^2 produced by the Sun on planets beyond the Earth is far too small to introduce any significant slowing of their rotation. The effects of the comparatively small satellites on the huge masses of the outer planets are also small enough to be ignored.

It would be reasonable, perhaps, to estimate a critical value of h^2 at approximately 2.0 as the limit consistent with habitability. (If anything, this is rather generous in view of the fact that the rotation of Venus has apparently been very greatly retarded by the Sun, despite a value of h^2 somewhat less than 2.0, but the uncertainties of the situation lure us to use the round figure as an approximation.) We will say, then, that for h^2 greater than 2.0, planetary rotation rates would probably be too slow (after the braking force had been working for the 3 billion years required for habitability) to be compatible with habitability.

Now we can return to this matter of minimum mass of a star consistent with its possessing habitable planets. For a habitable planet to possess the proper surface temperatures in the vicinity of a small main sequence star, it must orbit within an ecosphere placed quite close to the star—necessarily close since only then may enough illuminance be obtained from its feeble radiation. The closer such a planet must come to the star, however, the higher will h^2 be. There will come a point, as we consider smaller and smaller stars, where to come close enough to the star to receive enough illuminance for habitability means that the planet must also come close enough to have its period of rotation braked to the point of nonhabitability. In other words, for stars at the low-temperature end of the main sequence, the planetary temperature requirements for habitability are incompatible with rotation rate requirements.

A full ecosphere can exist around primaries with masses greater than 0.88 times that of the Sun. The entire ecosphere around such a primary is far enough out to be beyond any serious tidal braking effect on a planet's rotation. For primaries of lesser mass, the ecosphere is narrowed at its inner edge by the effects of increasing tidal braking. The ecosphere is narrowed to extinction when the primary's mass reaches a point as low as 0.72 times the mass of the Sun. For stars smaller than that, there is no orbit in which a planet can circle and find both comfortable temperatures and a bearable day-night cycle (except for rare cases of a planet-satellite system, to be taken up in the next section).

The range in mass of stars that could have habitable planets (without introducing satellite effects), then, is from 0.72 to 1.43 times the mass of the Sun. This corresponds to main sequence stars of spectral types F2 through K1.

Satellite Relationships

The rotation of a planet may be braked by a satellite as well as by its primary, as we have every reason to realize since the Earth's rotation is slowed by the Moon as well as by the Sun. Indeed, the Moon's braking effect on Earth is 5 times that of the Sun.

If the value of h^2 of a satellite on a planet is greater than 2.0, one would expect to find the planet's rotation halted with respect to the satellite after the 3 billion years that must have elapsed for the planet to have reached the habitable stage. (Since the Moon's braking effect on us is equivalent to an h^2 of 1.0, we have been slowed, but, even after possibly 5 billion years of braking, we have not yet been stopped.) Even after a planet's rotation has been stopped with respect to the satellite, it will continue with respect to the primary, and its solar day and synodic month will be of the same length.

The last situation is true for the Moon, for instance, as we have said before. Its rotation is stopped with respect to its companion body, the Earth, but continues with respect to the Sun. The Moon's solar day is 29½ Earthly days in length, and this is equal to the synodic month, the period of the Moon's revolution about the Earth from new moon to new moon.

Eventually, the Earth's rotation will be stopped with respect to the Moon, but by that time the Moon will have receded (thanks to the necessity of conserving angular momentum in a system in which the rotation of individual bodies is slowing) to the point where its period of revolution about the Earth will have lengthened to 55 24-hour days. Earth and Moon will then revolve, each presenting one face eternally to the other, and each nevertheless rotating with respect to the Sun. The length of the solar day for each body will be 55 24-hour days, equal to the length of the synodic month. Earth's year will then be 6½ days long.

For such a day-equals-month condition to be compatible with habitability, however, the period would have to be such as to produce a solar day less than 96 hours in duration—a figure that we have already chosen as the longest day consistent with habitability (see page 99). This means that the habitable planet and the satellite must not be too far apart. The farther apart, the longer the period of revolution of one about the other and the longer the resulting day.

Once the rotation of a planet with respect to the satellite has stopped, those tides on the planet due to the satellite will not be moving across the surface of the planet, but will remain fixed as a permanent bulge. There will still be tides due to the primary, of course. These, however, will not be braking the planet itself but the planet-satellite system as a whole, a much slower process, so that the further lengthening of the planet's day may be ignored.

If we consider cases of a stable planet-satellite situation in which higher and higher tides due to a primary are assumed, a new limiting condition will appear when the primary is massive enough or close enough to produce tides that reach a level so destructive as to be incompatible with land life. For instance, the erosion due to the power of the tides may become so excessively high that all the dry land on the planetary surface (assuming that any had formed in the first place) would disappear, leaving a continuous deep ocean swept twice daily by tides of enormous magnitude.

At what tidal magnitude would this occur? The Moon produces midocean tides on the Earth approximately 1 foot in height, yet local coastal tides may be much higher because of the piling up of water in shallow bays. It might be assumed, then, that midocean tides of the order of 10 to 20 feet would probably begin to be of sufficient magnitude to erode away all of the Earth's land masses over a period of many years. For present purposes let us assume that the destructive tide limit is indeed represented by midocean tides 20 feet in height.

Using this criterion, we can calculate all combinations of the tidal braking force due to a planet's primary and to its satellite. Such a calculation points to the existence of four types of planets: (1) those that are freely rotating with respect to both satellite and primary (as is the case with the Earth); (2) those with rotations halted with respect to a satellite but rotating freely with respect to the primary (as is the case with the Moon); (3) those like class 2, but where the tides due to the primary are large enough to be of destructive intensity (a class for which there are no known examples); and (4) those with rotations halted with respect to the primary (as is the case with Mercury, for instance, though Mercury, to be sure, is not a perfect example since it has no known satellite).

All habitable planets must be members of classes 1 and 2, and those in class 2 would be habitable only if their periods of rotation were less than 96 hours.

For a planet having the characteristics of

the Earth, the limitations on satellite mass and distance can be calculated. Where planetary bodies revolve about a mutual center of gravity, there is a minimum distance called "Roche's limit," after the astronomer who first worked it out. At a closer distance than this, the smaller of the two bodies would tend to break up into fragments as a result of the tide-raising forces of the larger. Saturn's rings, for instance, lie wholly within Roche's limit for that planet. This could mean that a once-innermost satellite of Saturn had broken up at some undetermined time in the past and that its fragments had spread out to fill the orbit. Or it could mean that material in the region had not been able to form a satellite by accretion in the face of nearby Saturn's strongly interfering gravitational influence. Jupiter's small innermost satellite, Amalthea (also called "Jupiter V" or "Barnard's Satellite") is almost at Roche's limit.

For the Earth, a small satellite would be inside Roche's limit if it were closer than about 10,000 miles. On the other hand, if such a satellite were farther off than 450,000 miles, it could not be retained in a near-circular orbit about the Earth. The Moon itself, with a mean distance of 239,000 miles from the Earth, falls just about midway between these two limits.

The ability of a satellite to halt a planet's rotation with respect to itself depends on both its mass and its distance from the planet. The smaller the satellite is, the closer it must be. The Moon

is of mass insufficient to have managed this so far
at its present distance. If the Earth possessed a
satellite at the minimum distance of 10,000 miles,
it could eventually stop Earth's rotation with respect to itself even if it were only one 10-thou-
sandth the mass of the Moon: a body, that is, only
a little over 100 miles in diameter, smaller than
some of the largest asteroids.

The important consideration, though, is
whether the planet's rotation rate, after having
been stopped with respect to the satellite, is consistent with habitability. In order for this to be
so, the synodic month must be less than 96 hours,
and this can only be so if the satellite is closer than
65,000 miles at the time the rotation is stopped.
Our Moon is far more distant than 65,000 miles,
and when Earth's rotation is stopped with respect
to it, its day will be far longer than 96 hours, and
Earth will then be non-habitable. In fact, Earth
will become non-habitable long before its rotation
with respect to the Moon has been completely
halted.

If the satellite is comparable in mass to the
planet itself, there is the possibility that it, too,
will be habitable. We will then have twin habitable planets. The dimensions of Roche's limit
rise somewhat as the satellite increases in mass,
and so does the maximum distance allowed for a
month of 96 hours or less. Such twin habitable
planets would have to be separated by a distance

of not less than 18,000 miles and not more than 100,000 miles.

It is possible, of course, that a habitable planet may have a satellite larger than itself. For example, if Jupiter revolved in Earth's orbit and Earth revolved about Jupiter, Earth could still be a "habitable planet" as defined in this book and, from an Earth-centered viewpoint, Jupiter would be its giant satellite. (In view of the usual feeling that a satellite is the smaller of two related non-luminous bodies, it might be better if we referred to such a giant satellite by the more neutral word "companion.")

But now a new type of limit could be conceived of, one determined by the heat produced by the more massive body. As a giant companion approaches stellar mass, its surface presumably becomes hotter because it has not been able to lose its heat of gravitational accretion as quickly as a small planet might, and because it is also receiving heat from the primary. At some mass level, a companion, even though it is not hot enough to maintain thermonuclear reactions in its interior and become a star, is nevertheless hot enough to cause the loss of water from the atmosphere of the otherwise habitable planet revolving about it, making it no longer habitable. Since we have little knowledge of the surface temperatures of bodies of mass greater than Jupiter, we can only locate the upper limit for the size of the companion rather approximately, and place it at the point where its mass is 10 times that of Jupiter.

A companion of this mass would have to be at least 150,000 miles from the Earth, or Earth would be ripped apart by the tidal forces produced by its giant neighbor. (In other words, Earth would be within the Roche's limit of the companion.) Nor could the separation be much over a million miles, or the length of the month would exceed 96 hours.

For a planet unescorted by a satellite large enough or close enough to stop its rotation and maintain it at that point despite the tides produced by the primary, the minimum mass for the star was set at 0.72 times the mass of the Sun. If, however, a planet *does* have a satellite that can slow its rotation with respect to the Sun to a period of not more than 96 hours and then maintain that period, that planet could crowd into an ecosphere very close to the primary. Stars of unusually small mass could therefore serve as primaries for such planets, although they could not serve as primaries for habitable planets without appropriate satellites. In fact, the new minimum limit for the mass of a primary comes only at a point where the planet must crowd so close to the primary that tides become destructively high.

The new lower limit on stellar mass for the special planet-satellite system just described would be only half the one originally given, or about 0.35

times the mass of our Sun. Stars of spectral classes from K2 down to M2 might serve as primaries for such planet-satellite combinations. It must be admitted, though, that planet-satellite combinations having just the right set of properties are probably very rare. We will therefore continue to consider 0.72 times the mass of our Sun as the minimum mass for the primary of a habitable planet.

Special Properties of Binary Star Systems

So far, we have been concerned with planets revolving around isolated stars. A large fraction of all stars, however, exist in multiple stellar systems—double, triple, and quadruple. The most common type is the double star, or "binary," system, made up of two stars circling about a mutual center of gravity. In the minority of cases where more than two stars are gravitationally associated, the other members are far removed from the first two—so far removed that they can have little effect on planets belonging to the first two. Thus, if we examine the special properties of binary star systems important to habitable planets, after having already considered the properties of planets about isolated stars, we will have covered all the classes of stars that are important for the purpose of this book.

Two essential questions must be analyzed with respect to the existence of habitable planets in binary star systems: (1) Can stable planetary orbits exist at the proper distances from a star in a binary star system? (2) If so, is the illuminance at the planet (the sum of the radiation received from both stars) constant enough to be consistent with habitability?

Some of the factors that must be considered in the most general form of the problem of a binary star system are the spectral types and masses of the two stars, their average distance of separation, the eccentricity of their orbits, the inclination of the plane of the planetary orbit to the plane of the orbiting stars, and so on. Because of the vast number of possible combinations of these astronomical factors, the problem must be simplified before analysis is practical. At the outset, then, let us limit the problem to the simplest possible case: one in which the two stars move in near-circular orbits around their common center of gravity and the planet is in a near-circular orbit around one or both of the stars, in the same plane as the two orbiting stars.

In this case, two types of stable orbits can exist. A planet can circle in the near neighborhood of one of the stars at such a distance that the gravitational influence of the primary is paramount, while that of the comparatively distant companion star is small enough to be ignored. Even the less massive star of a double star system can usually maintain a stable, near-circular planetary orbit against the influence of its more massive partner, provided the two stars are separated by a considerable distance. Naturally, the less massive a particular star is, the closer the planetary orbit must be to the star if it is to achieve stability.

A second type of stable orbit can exist at a comparatively great distance from the two stars, a distance great enough to enable the planet to circle the center of gravity of the two stars, with the separate gravitational fields of the stars themselves too weak to perturb the planetary orbit seriously. The two stars, in other words, would in this case serve as a single primary divided into two parts.

Thus, the first requirement for habitability in a double star system—the existence of stable near-circular orbits—is fulfilled.

If at least one of the regions in which stable planetary orbits can exist also includes an ecosphere, and if, within that ecosphere, the total illuminance is fairly constant at all points in the orbit for all arrangements of the two stars, then the second requirement is fulfilled, and habitable planets can exist in a binary star system.

How much variability in the level of illumination would be permissible? This is a very difficult problem to assess because of the inadequacy of our knowledge concerning the prediction of planetary surface temperatures from information about stellar radiation, and because the possible combinations of planetary variables are so numerous. However, let us assume (perhaps over-conserva-

tively) that the permissible variation in total illuminance as a result of having two sources of radiation for any given planet in a near-circular orbit
must be less than 10 per cent during the annual
period. (The variation is about 7 per cent for the
Earth because of the eccentricity of its orbit.)

It turns out that in the case we are considering, there are regions in the immediate vicinity of
each star where the illumination contours are
nearly circular. This means that a planet on a
near-circular orbit within these regions would receive a nearly constant supply of radiation, virtually all of it coming from the primary and comparatively little from the companion star.

At a great distance from both stars, the illumination contours are again nearly circular, with
the more massive and brighter star at the center.
A planet in a distant near-circular orbit around
both stars would move about the center of gravity
of the two stars; hence it would tend to cut across
illumination contours (and receive varying illuminance) unless it were quite far away. If it were
not far enough away, in other words, it would receive distinctly less illuminance when the smaller
star was between it and the larger star, and distinctly more illuminance when the larger star was
between it and the smaller star.

To take a specific case, imagine two orbiting
stars separated by a distance that we can arbitrar-
ily set equal to 1.00. Let us further suppose that

the brighter star is 25 times as bright as the dimmer one and about 2.3 times as massive. If we require planetary orbits to be in one of the stable regions, with the variability of illuminance not exceeding 10 per cent, we can calculate as follows: Around the more luminous star, the orbit must lie inside a circle of radius 0.319, while around the less luminous star it must lie inside a circle of radius 0.108. Finally, a distant orbit around both stars must lie outside a circle of radius equal to 10.5.

To convert this into actual miles, let us suppose that the two stars are separated by 2 billion miles. (This is a little more than the distance separating the Sun and Uranus, and it is about equal to the average distance separating Alpha Centauri A and Alpha Centauri B, the system mentioned at the very beginning of the book.)

A stable near-circular orbit with reasonably constant illumination could then exist within 640 million miles of the brighter star (Jupiter falls well within that limit with respect to the Sun) and within 220 million miles of the dimmer star (Mars falls within that limit with respect to the Sun). If both stars lay in the permitted mass range, then each would be able to possess habitable planets, since the ecosphere in both cases would lie inside the limits just cited. (The dimmer star could only support a planet with a satellite capable of maintaining its rotation.)

Another set of stable near-circular orbits would carry a planet about both stars as a gravitational unit, at a distance of not less than 21 billion miles, or 5 times the average distance of Pluto from the Sun. For a star system like the one under discussion (very much like the Alpha Centauri system), such orbits would be far outside the ecospheres and could not contain habitable planets. However, if the orbiting stars were very close together (within a few million miles of each other, as is true of a common class of double stars called "spectroscopic binaries"), then it could be the distant orbit circling both stars that would lie within the ecosphere. Planets orbiting around one or another of the two closely spaced stars of a spectroscopic binary would be unlikely indeed, and would, in any case, not be habitable.

We conclude, then, that if the stars of a binary system are separated by a sizable distance in the billions-of-miles range, then either or both might conceivably possess a habitable planet in orbit about itself. If the stars of a binary system are separated by a distance in the low millions-of-miles range, then both can serve as a common primary for a habitable planet circling their center of gravity. There will be binary systems with intermediate separations such as to preclude the existence of ecospheres either about the stars singly, or about the stars taken together, but these cases are likely to be comparatively rare.

Summary of Requirements for a Habitable Planet

Its *mass* must be greater than 0.4 times the mass of the Earth to permit it to produce and retain a breathable atmosphere, and less than 2.35 times the mass of the Earth since surface gravity must be less than 1.5 g.

Its *period of rotation* must be less than about 96 hours (4 Earth days) to prevent excessively high daytime temperatures and excessively low nighttime temperatures.

The *age* of the planet (and the star around which it orbits) must be greater than about 3 billion years to allow for the appearance of complex life forms and the production of a breathable atmosphere.

The *illuminance* at low *equatorial inclination* should lie between 0.65 and 1.35 times Earth normal to maintain temperatures consistent with habitability, although certain combinations of illuminance and equatorial inclination, up to 1.9 times Earth normal for the former and up to 80° for the latter will allow marginal habitability.

The *orbital eccentricity* must be less than approximately 0.2, since greater values might produce unacceptably extreme temperature patterns on the planetary surface.

The *mass of the primary* must be less than 1.43 times the mass of our Sun, since only then will its residence time on the main sequence be greater than 3 billion years. For habitable planets in general, the mass of the primary must be greater than 0.72 times the mass of our Sun because, when primaries have masses smaller than this, there is an incompatibility between permissible illuminance and permissible rotation rate. For a special (probably quite rare) class of planets with extremely large or close satellites, there is an extension of the lowest permissible primary mass down to 0.35 times that of our Sun.

If the planet orbits in a *binary star system,* the two stars must either be quite close together or quite far apart for ecospheres to exist in a region where stable orbits are possible.

If all of these requisites are satisfied, then there is a very good possibility that a planet will indeed be habitable.

Probability of

Habitable Planets

Having summarized the properties of habitable planets and the astronomical factors required to produce these properties, we can now attempt to estimate the numbers of such bodies in our own Galaxy (the Milky Way). To do this with some reasonable degree of accuracy (in the liberal spirit of this book), it is necessary to consider the following factors:

1. The prevalence of stars in the suitable mass range—that is, 0.35 to 1.43 times the mass of our Sun—and, in particular, the prevalence of stars in the mass range from 0.72 to 1.43 times the mass of our Sun.

2. The probability that a given star has planets in orbit around it.

3. The probability that the inclination of the

planet's equator is correct for its orbital distance.

4. The probability that at least one planet orbits within an ecosphere.

5. The probability that the planet has a suitable mass—that is, 0.4 to 2.35 times the mass of the Earth.

6. The probability that the planet's orbital eccentricity is sufficiently low—that is, less than 0.2.

7. The probability that the presence of a second star has not rendered the planet nonhabitable.

8. The probability that the planet's rate of rotation is neither too fast nor too slow—that is, with a length of day between 3 hours and 96 hours.

9. The probability that the planet is of the proper age—that is, over 3 billion years old.

10. The probability that, all astronomical conditions being proper, life has developed on the planet.

Once quantities for all these factors have been established, the probability that a given star of appropriate mass range may have a habitable planet can be obtained by multiplying all the factors from 2 to 10, inclusive. The total number of habitable planets in our Galaxy can then be obtained by multiplying this product by factor 1.

Obviously, the final number is bound to be highly uncertain, since not all of the above factors are known with accuracy. In fact, most of them can only be estimated rather roughly; and the values assigned to some of them depend on which theory of the structure and origin of planets one happens to favor. The estimates contained in this chapter and the next are not, therefore, intended to represent the final word on the subject, but are merely the best we can do to arrive at numbers that have at least some rationale behind them. Let us then deal with each factor separately.

Factor 1: Stars of Suitable Mass

Counts of the numbers of stars of the various magnitudes and spectral types in the neighborhood of the Sun have been made by astronomers. From these counts and from the relationships between magnitude, mass, and spectral type, it is possible to estimate the concentration of stars of each spectral type in the solar neighborhood. (We have no choice but to assume that the distribution of stars in the solar neighborhood is typical of the Galaxy as a whole.)

For instance, out of every 10,000 stars, it is estimated that 9923 belong to one part or another of the main sequence (the stable stage of a star's evolutionary development). The remaining 77, outside the main sequence, include among them the red giants, such as Betelgeuse and Antares; the white dwarfs, such as the companions of Sirius and Procyon; and so on. The stars outside the main sequence are not suitable for the possession of habitable planets, but this scarcely makes a dent in the vast numbers of stars.

Among the main sequence stars, the least massive and least luminous are the most common. At each spectrum level marking an increase in mass and luminosity, the number of representatives decreases. Out of every 10,000 stars, it is estimated that 7325 are of spectral class M on the main sequence. In other words, this class, which includes the dimmest and least massive stars, makes up nearly three-fourths of all stars. Moving up systematically through the spectral classes, 1509 are of spectral class K; 731 are of spectral class G; 291 are of spectral class F; 58 are of spectral class A; and 9 are of spectral class B. The one remaining spectral class is O, making up the most massive, most luminous, and hottest of the stars of the main sequence. Representatives of this spectral class are very rare; only one star out of 50,000 is a member.

The more massive and luminous stars, although relatively rare, are extremely prominent in the night sky since, because of their brightness, we can see all of them out to a great distance. On the other hand, the less massive and less luminous stars, although much more common, can only be seen if particularly close to our Sun, and even then they are likely to be inconspicuous.

Of the 20 brightest stars in the sky (those of

tion. If we wish to restrict ourselves to those stars with reasonable probabilities of having a habitable planet, we might eliminate this lower extension of the range, which would extend from spectral classes K2 to M2. This would leave us with only the range of spectral classes from F2 to K1. Even this is not extremely restrictive, for these stars make up 13 per cent of all stars.

It will also prove useful to consider the prevalence of stars in each spectral class within the range F2 to K1. Naturally, the stars of spectral class F2 are the least numerous (0.32 per cent of all stars), while those of spectral class K1 are the most numerous (1.1 per cent of all stars).

Putting this in fractional notation, the stars of spectral classes F2 to M2, inclusive, make up about 1/4 of all the stars; while stars of classes F2 to K1, inclusive, make up about 1/8 of all the stars. Within the latter range, the prevalence of stars in each spectral class ranges from 1/300 for F2 to 1/90 for K1.

The total number of stars in our Galaxy has been estimated at 135 billion. Therefore, the total number of stars in the range of spectral classes from F2 to M2, representing all stars that might conceivably have habitable planets, is 34 billion.

The total number of stars in the range of spectral classes F2 to K1, which can have habitable planets without special satellites (and this is the important range), is 17.5 billion.

The total number of stars by spectral class ranges from 0.45 billion for spectral class F2 up to 1.5 billion for spectral class K1.

In sheer numbers, then, there is no lack of suitable stars in the Galaxy.

Factor 2: Stars with Planets

Ideas about the formation of the Solar system prevalent in the first third of this century (and which are still current in some quarters) depend on a catastrophic origin—a near collision (or even an actual one) between the Sun and a passing star. This hypothesis requires the planets to have been formed from incandescent material pulled out of one or both of the two stars as they passed each other.

A theoretical analysis of this situation makes it appear quite improbable that any material so drawn out would remain in orbit after such an encounter; it would instead tend to fall back into the two bodies once they had separated. Moreover, even if such incandescent material did remain in orbit, it would have no tendency to coalesce into the form of large droplets, as has often been suggested. (At the time when the tidal theory of planetary formation was advanced, it was not fully realized how incredibly hot the interiors of stars are.) Substance from stellar interiors dragged out by tidal influence would, in view of its temperature, literally explode into its constituent atoms or even into subatomic particles. The end result, at best, would be a star surrounded by a flat ring of gas and dust. (Within this ring, planets might subsequently form by accretion, but this is not the usual chain of events hypothesized in catastrophic theories of planetary formation.)

An important consequence of the tidal-filament, near-collision mode of planetary system formation is that such systems would be rare in the Galaxy. Individual stars of the Galaxy are separated by distances so large in comparison to their diameters and their velocity of travel that near collisions among them could only be expected at tremendously long intervals. In consequence, there might be no more than a handful of planetary systems developed in the Galaxy throughout all its long history. It might then be very unreasonable to expect that more than one planetary system of these very few would contain a habitable planet on which any form of complicated life could develop, and that one system would, naturally, be our own. Mankind, according to that view, would very likely be alone in the Galaxy and would have small hope of finding anything in his interstellar travels but stars and empty space and more stars and still more empty space.

However, the catastrophic theories are no longer accepted by astronomers generally. The theory of planetary formation by accretion (which is now widely accepted) implies that planetary systems are the rule rather than the exception. Some small fraction of the total mass of the original cloud out of which the star was formed should almost surely have a velocity relative to the central star sufficiently high to prevent its being captured by the star. Out of this, the planets would be formed.

For that reason we can assume that on the basis of currently held ideas about the formation of stars within clouds of dust and gas, it is reasonable to suppose that *every* star (and certainly every star in the mass range that interests us) has a family of non-self-luminous bodies in orbit around it.

We will therefore take the probability that a given star has planets in orbit around it as 1.0.

Factor 3: Correct Equatorial Inclination

As we have seen, the surface temperature pattern of a planet depends primarily on its average distance from its primary and on the inclination of its equatorial plane to the plane of the planet's orbit about its primary. A fairly complex relationship exists between habitability, mass of primary, mass of planet, and mean distance from primary with respect to the braking of rotation due to tidal effect at the inner edge of the ecosphere. For example, large planets, when associated with small primaries, would have wider ecospheres than small planets would have when associated with the same primary. This is because the large planets, having higher rotational energies, are not so rapidly retarded.

Because of the complexity of the relationship, it would be impractically complicated to compute the probability of planet habitability for all possible combinations of inclination and illuminance, together with all possible combinations of mass of primary, mass of planet, and distance of primary-planet separation. In this book, however, the emphasis is not so much on absolute precision as on obtaining certain rough but as-reasonable-as-might-be-expected estimates; some simplification is therefore desirable.

We can begin by considering the equatorial inclinations of the 6 planets of the Solar system for which the value is known with precision. (As we stated near the beginning of the book, we have decided to take the planets of the Solar system to be a typical assemblage of planets and as representative of all the stellar systems that exist unobserved by us.) Of these 6 planets, one (Jupiter) has an unusually low inclination, $3°$, while another (Uranus) has an unusually high one, $98°$. The equatorial inclination of the remaining 4 (those of the Earth, Mars, Saturn, and Neptune) all lie between $20°$ and $30°$, with that of Earth being $23.5°$.

An analysis into which several simplifying assumptions were introduced would make it appear that there is a probability of 0.81 that the equatorial inclination would be less than 54°, a value compatible with habitability for large portions of the planet. We will let that probability stand over the whole range of tolerable illuminance, as an approximation, rather than try to calculate how it varies for different levels of illumination. This greatly simplifies the subsequent calculations without, we hope, introducing any great errors.

It will be assumed, then, that the probability of suitable inclination is 0.81.

Factor 4: Planet within the Ecosphere

As explained earlier (see page 124), primaries having masses greater than 0.88 times that of our Sun have complete ecospheres, while those with masses between 0.72 and 0.88 times that of our Sun have narrowed ecospheres because of the rotation-retarding effects of the primary at the inner edge of the ecosphere.

With that reminder, we go on to estimate the probability that there will be at least one planet orbiting within an ecosphere. The planets in the one planetary system that we can study in detail (our own) are spaced in such a way that each, with the exception of the atypical planet Pluto, falls outside the forbidden regions of its neighbors (see page 86). The space of the Solar system is one half taken (approximately) by the planets and their forbidden regions, spaced in such a way that there are comfortable gaps between them.

We can assume that this pattern may well be a universal feature of planetary systems around other stars.

We can begin, then, by computing the probability that at least one planet orbits in a given distance interval within our own Solar system. Suppose we take, for instance, a distance interval such that the outer rim of the interval is 3.41 times as far from the Sun as the inner rim. Such an interval placed anywhere within the Solar system (provided the inner rim is not beyond the orbit of Pluto and the outer rim is not inside the orbit of Mercury) is bound to contain the orbit of at least one planet. This is because the widest interplanetary spacing in the system, in proportion to total distance, is that between Jupiter and Mars, and Jupiter's orbit is just 3.41 times as distant from the Sun as is that of Mars.

For the interval ratio of 3.41, then, the probability of finding a planet within the distance interval is 1.0. For narrower distance intervals, the probability of finding a planet's orbit within that interval is less than 1.0, and the value can be calculated from data based on the structure of our own Solar system.

We have set the ecosphere as lying between illuminances of 1.35 times Earth normal at the edge closest to the primary, to 0.65 times Earth normal at the edge farthest from the primary. To translate this into actual distances, we must remember that illumination varies with the square of the distance. Keeping this in mind, it turns out that the outer edge of a full ecosphere is 1.441 times as far from the primary as the inner edge is. The probability that at least one planet orbits within such a distance interval is calculated as 0.63. For primaries with masses below 0.88 times the mass of our Sun, the ecosphere is narrowed at the inner edge; and with a narrowed ecosphere, the probability of the existence of a planet within the interval is correspondingly reduced.

This means that the probability of finding a planet within the ecosphere of a given star varies with its mass and, therefore, with its spectral class. Stars of spectral classes F2 to G4 (within the permitted range for habitability) have full ecospheres. (Our Sun, nominally of class G0, therefore has a full ecosphere—though this may be a marginal affair, since Venus, near the inner edge of the ecosphere, seems to have had its rotation severely retarded by the Sun's tidal action.) For such stars, the probability of having a planet is 0.63.

For stars of spectral class G5, where the narrowing of the ecosphere becomes evident, the probability is down to 0.60. The narrowing grows steadily more marked as we make our way down the spectral classes until, by the time spectral class K2 is reached, it is down to 0.0.

To be sure there is also the possibility that two planets, in independent orbits, may circle within the ecosphere of a star. If Venus, for example, were 13 million miles more distant from the

Sun than it is (as it could just as well have been for anything we know to the contrary—there would be no interference with Earth's forbidden region), our Solar system would have had two independent planets, Earth and Venus, within its ecosphere. The probability of that happening in the general case, however, is calculated as only 0.036. The probability of two planets occurring within the ecosphere in the form of twins, circling each other, is even lower. Such cases can be ignored, we think, without introducing much error.

We conclude, then, that for stars in the suitable mass range, the probability that at least one planet is orbiting within the ecosphere varies with spectral class, being 0.63 for spectral classes F0 to G4 and declining progressively thereafter to 0.0 for spectral class K2.

Factor 5: Planet of Suitable Mass

The masses of the planets of the Solar system exhibit a pattern of regularity. The smaller individuals among the planets (remember that our definition of "planet" includes a number of the bodies ordinarily called "satellites") are more numerous than the larger ones. This connection between "smallness" and "more numerous" is also found to be true among the stars, the asteroids, and the meteoroids, so that we can feel safe in assuming such a relationship to be a significant one, and not accidental.

Under the assumption that the planets of our Solar system represent a fair sampling of planets in general, we can construct a probability distribution pattern for the masses of planets. Such a distribution pattern would make it appear that the probability of a planet, chosen at random, lying within the mass range 0.4 to 2.35 times the mass of the Earth (the range compatible with habitability) is equal to 0.19.

From one point of view, this may appear to be an unnecessarily pessimistic approach since, in our system, there is a rough, though by no means regular, variation of planetary mass with distance from the Sun. The small terrestrial planets, near the suitable mass range, are close in, near the ecosphere; the massive giant planets, far too large for habitability, are also far distant from the Sun and hence far beyond the ecosphere. This may suggest a general law that could be applied to all systems, to the effect that planets near the habitable range of mass are bound to be found near the ecosphere. If so, the chances of a planet of habitable mass being found within the ecosphere should be considerably higher than 0.19.

However, since there are so many irregularities in the mass-distance pattern, and since attempts to explain it in detail have not been very successful, we have decided to adopt the more general and conservative idea that planetary mass is independent of distance from the primary. We will therefore take the probability of a planet having a suitable mass as 0.19.

Factor 6: Correct Orbital Eccentricity

An inspection of the values of orbital eccentricity among the bodies of planetary size in the Solar system makes it possible to work out an empirical system for calculating the probability that the eccentricity is less than any assigned value. By this system, if we permit orbital eccentricities up to 0.2, then about 94 per cent of the planets should have eccentricities below this value and thus should be habitable, provided all the other essential requirements are met. This is not surprising, considering how low the orbital eccentricities are for almost all the planetary bodies in the Solar system. It seems only reasonable to suppose this will hold true for other stellar systems as well.

We will take the probability of a suitable orbital eccentricity, then, to be 0.94.

Factor 7: Non-interfering Companion Star

As discussed earlier (see page 134), habitable planets can exist in binary star systems if the two stars are so close together that there is a single ecosphere around the pair or if they are so far apart that at least one can have an ecosphere without interference from the other. In either case, the existence of a companion star does not render the planet uninhabitable. Furthermore, if a third and even a fourth star are also part of the system, they will be so far removed from the pair that it can be assumed they will not interfere with the habitability of the planet.

The proportion of stars that are spectroscopic doubles (binaries so closely spaced that they give evidence of their separate existence only through spectroscopic effects) has been estimated for different spectral classes. These estimations are based on actual star-counts, but the conclusions reached vary widely just the same. In spectral classes F, G, and K, the proportion is 4 to 7 per cent according to one authority, and 28 to 30 per cent according to another. However, the periods of revolution of spectroscopic doubles made up of main sequence stars are typically measured in days. This means that the two stars must be quite close together, with separations, for the most part, of less than 10 million miles (sometimes considerably less). Such close spacing would be quite compatible with the existence of an ecosphere around both stars taken as a unit.

In addition, there are visual doubles (those binary stars spaced far enough apart to be seen separately in a telescope). Of these, 12 per cent have measurable orbital motions, and for them the separations are already great enough to permit the existence of a suitable ecosphere around either star. The remaining visual binaries, with unmeasured orbital elements, are, in all likelihood, even more widely separated; and each of the pair is even less likely to interfere with planets about the other. Some 4 to 9 per cent of the stars in spectral classes F, G, and K are visual doubles, and probably no more than a fraction of 1 per cent of those stars are visual doubles that have separations in the critical distance range that would prevent the existence of an ecosphere.

It seems to us (contrary to the opinion of some astronomers, it must be admitted) that practically all the spectroscopic doubles could possess habitable planets, and that among visual doubles, spacings that would prevent the existence of a normal ecosphere are rare.

In view of the incomplete state of our knowledge, however, let us estimate that generally, for any star taken at random, there will be interference due to a companion star in 5 per cent of the cases, to allow for possible underestimates of the numbers of binary star systems having awkward separation distances.

For stars generally, then, the probability that a companion star will not interfere with the habitability of a planet is 0.95. (For stars known to be isolated or to have companions at non-awkward distances, the probability is, of course, 1.00.)

Factor 8: Correct Rate of Rotation

As we have seen in Chapter 3, there seems to be a definite relationship between planetary mass and rate of rotation for those planets of our Solar system that have unrestricted rotation. The greater the mass, in other words, the shorter the period of rotation. Similar relationships may well be found in other planetary systems, since the same evolutionary and dynamic forces would, we confidently expect, be working in a similar manner.

Based on this assumption, planets having masses close to that of the Earth would also have periods of rotation close to that of the Earth— or somewhat faster, considering that the Earth's rotation period has been slowed by the Moon (see page 125)—except where their rotation is retarded strongly by tidal effects from satellites or primaries. It would seem, then, that we could make the general assumption that a planet with a mass suitable for habitability would also have a rotation suitable for habitability. Because of the sparseness of the data, however, and because we do not have much confidence that we know all the factors that determine a planet's rotation rate, let us consider the probability of a suitable rotation period to be only 0.9 for tidally unretarded planets.

(Again we might mention that habitable close-in planets with satellites large enough or close enough to produce a day-month in the suitable range, after the fashion described on page 130, form an added group; but again we might point out that this is a very small group and need not be considered as affecting our calculations.)

Factor 9: Correct Age

We have concluded previously (see page 106) that it takes something of the order of 3 billion years to produce a habitable planet, provided all the astronomical conditions are correct. For a star to have a habitable planet, then, it must have been delivering radiation at a reasonably constant rate for at least 3 billion years.

We know (or believe that we know at least approximately) the relationship between stellar mass and time of residence in the main sequence (where radiation is reasonably constant). Knowing this, we can determine the maximum age for a main sequence star of given mass. It cannot be older than its total life expectancy on the main sequence or it would not be on the main sequence any more. It also probably cannot be older than the galaxy in which it exists.

It is not so easy to determine the actual age of a star of given mass. The minimum age could be as low as zero, for we know that stars are still being formed in our Galaxy. There is not, at present, a reliable way of determining whether a given main sequence star was formed fairly recently or whether it is nearing the end of its residence on the main sequence.

If we assume that stars are forming and have been formed during the past at a fairly constant rate, however, we can calculate the probability that a given star is older than, say, 3 billion years, basing the analysis on the main sequence lifetime of stars of a particular mass and luminosity and on the age of the galaxy in which the star is to be found. We must also assume that planetary systems are formed concurrently with their primary stars, so that planets and primary will be of approximately the same age.

Based on different bits of evidence and different interpretations of the observational data, current estimates of the age of our own Galaxy range from 10 to 25 billion years. If we take 10 billion years as our base of calculation, so as not to overestimate the probabilities, then the likelihood that a given star is older than 3 billion years varies with the spectral class. The stars of spectral classes F0 and F1 stay so short a time in the main sequence that an age of 3 billion years for any one of them that is still in the main sequence is extremely unlikely. Only beginning with spectral class F2 can a 3-billion-year lifetime in the main sequence become possible, and the probability of such an age for any given star rises as one goes down the spectral classes until a maximum of 0.70 is attained for spectral class G0 and beyond.

Factor 10: Development of Life

As we have seen, free oxygen in a planetary atmosphere is essential for a planet to be considered habitable. It is highly probable that virtually all the free oxygen in the Earth's atmosphere has come from the decomposition of water by green plants during photosynthesis. What are the chances, then, that a similar development will take place on other planets that are astronomically suitable for life?

We do not actually know how life started on the Earth, although the topic has been much discussed in recent years. If we assume that the origin of life has been a natural evolutionary process, however, there is good reason to suppose that life would always originate whenever the conditions were correct. In support of this view is the evidence that microscopic forms of life appeared on the Earth as soon after the Earth's formation as would seem possible. The oldest rocks, on the basis of analyses for radioactive decay products, are estimated to be about 4.5 billion years old (though the planet is probably somewhat older than that), while the earliest detectable traces of life forms appear in rocks that are about 2.5 billion years old. Since the appearance of life forms that we know about must have been preceded by a long period of the development of chemical products of gradually increasing complexity and of simple life forms that happened to leave no permanent evidence of their presence, it would seem that life originated on the Earth just about as soon as the environmental conditions permitted.

Since living matter, as we know it, is composed of some of the most abundant elements in the Universe and in the Earth's crust, it seems reasonable to suppose that life on habitable planets generally would be built out of the same abundant elements and be similar to Earth life in general chemical composition, although undoubtedly differing from Earth life in detail. After all, wherever they are found, living organisms must always depend on the same basic chemical processes and physical laws with which we are familiar on the Earth's surface. If water is the only major source of hydrogen available (as is very likely to be true), then living material, wherever it exists, will depend on the development of some process for extracting hydrogen from water and incidentally releasing oxygen to the atmosphere.

Life, wherever it appears, may depend on very similar organic compounds for the following reasons. Water, ammonia, and methane (which are compounds of hydrogen with oxygen, nitrogen, and carbon, respectively) should be among the most abundant compounds in the primitive atmospheres of terrestrial planets in the early stages of their development. A variety of more complicated organic compounds are formed when mixtures of methane, ammonia, and water are acted upon by electrical discharges or by various kinds of radiation—a fact borne out by actual experiments. Included among these compounds are the simple building blocks of the proteins and nucleic acids, the complex substances on which living tissue is based. Given the same starting materials, then, and the same building blocks produced from them, it seems reasonable to suppose that similar complex final products ought to be developed, although the final products on each planet may differ in important detail from those on every other planet.

One of the atom combinations that ought to be built up readily from methane, ammonia, water, and hydrogen is the porphyrin ring. This is a stable substance that forms the basic skeleton of the structure of chlorophyll and also of heme, the key component of hemoglobin, which is, in turn, the red, oxygen-carrying protein of the red blood cells. It is not in the least unlikely, then, that the chemical similarities of life forms developing separately on different planets may extend to the actual chemical groupings used to catalyze photosynthesis in plants and oxygen transport in animals (though, again, there are sure to be differences in detail).

Of course, the fact that life develops on other planets and follows the general chemical line of life on Earth still does not absolutely ensure that such extraterrestrial life forms would be palatable, or even edible, from the standpoint of man or animals that have evolved on the Earth. Particular vitamins may be absent, or compounds toxic to us may be present. The tissues of alien life, even if not poisonous, may be indigestible or simply

foul-tasting. (There are, after all, numerous forms
of life native to Earth itself that cannot be eaten
by man.)

Space colonists, then, will certainly be prepared to seed an otherwise habitable planet with
the plant life, soil bacteria, pollinating insects,
and so on, of the Earth, to eke out the marginal
food supply (if any) offered by the indigenous
life and to supply the accustomed diet of foods
that they prefer, and that provide the particular
amino acids and vitamins needed by human beings. (It is quite probable, incidentally, that life
forms that have evolved on the Earth will encounter no natural parasites or disease-producing
agents on an alien planet, since parasites and the
life forms on which they live must evolve together. It would therefore be worth making an
effort to bring as few parasites as possible to a
new colony.)

The probability that native *intelligent* life
will be present on a given planet may be quite
low. There are enormous numbers of possible
paths that evolution can take, and, based on the
past history of the Earth, few of them necessarily
lead to high intelligence. On the Earth, it has
been only within the past few hundred thousand
years or so (out of the billions of years during
which living things have existed) that the presence of an intelligent species would have been
apparent to a visitor from some other planet. It
may be that given enough time, an intelligent

species would eventually appear on any given habitable planet. However, its time of appearance is probably not highly predictable. If, through some unlucky accident, the ancestral family group of the human race had become extinct so that there were now no human beings on the Earth, how long might it be before some other intelligent land species would evolve from the existing animal species of the Earth? The answer is by no means evident.

On the other hand, once an intelligent species has evolved on a planet, there is a good chance that it will quickly spread to other unoccupied habitable planets in its region of space, as the human race may well do within the span of not too many generations.

In view of that possibility, the chance of finding a habitable planet already in the possession of an intelligent species and finding ourselves, as we spread outward, in a face-to-face confrontation with another species, also spreading outward, is one of the more interesting things that we have to look forward to. In assessing the probability of this, however, we have so little to go on, that for present purposes, the matter will not be considered in making our calculations.

We will satisfy ourselves, then, with assuming that life (very likely non-intelligent) will always appear on planets having the correct combination of astronomical conditions and that free oxygen in the atmosphere will always accompany the appearance of life on a large scale. The probability assigned to this factor is 1.0.

Probability and Total
Number of Habitable Planets

Let us now list the probabilities of factors 2 to 10 inclusive:

Factor	*Probability*
2. Stars with planets	1.00
3. Correct equatorial inclination	0.81
4. Planet within the ecosphere	0.00 to 0.63
5. Planet of suitable mass	0.19
6. Correct orbital eccentricity	0.94
7. Non-interfering companion star	0.95
8. Correct rate of rotation	0.90
9. Correct age	0.00 to 0.70
10. Development of life	1.00

For factors 4 and 9, the probabilities vary with the spectral class. Factor 4 sets a minimum spectral class of K1 (except for a rare extension to M2 for satellite-stabilized planets). Factor 9 sets a maximum spectral class of F2.

The probability that a given star will contain at least one habitable planet in orbit around it is calculated by multiplying all the probabilities of the factors listed above. Because of the variability with spectral class of the values for factors 4 and 9, it is necessary to make the calculation for each spectral class separately. In order to do this, the exact values of factors 4 and 9 for each spectral class must be used. Without listing these values here, we will nevertheless use them, and list only the final results.

The following, then, are the probabilities that a star of a particular spectral class will possess a habitable planet in orbit about it:

Spectral Class	Probability	Spectral Class	Probability
F2	0.0106	G2	0.0545
F3	0.0192	G3	0.0545
F4	0.0303	G4	0.0545
F5	0.0344	G5	0.0520
F6	0.0418	G6	0.0511
F7	0.0457	G7	0.0394
F8	0.0499	G8	0.0312
F9	0.0529	G9	0.0221
Go	0.0545	Ko	0.0147
G1	0.0545	K1	0.0061

The highest probability comes in the range Go to G4. A star in these spectral classes, taken at random, has a 0.0545 probability (1 chance in 18) of having a habitable planet in orbit around itself. It is in precisely this range that our own Sun (spectral class Go or G2) falls, and the range represents one of the stars with masses from 0.89 to 1.04 times that of our Sun.

The probability that a star will have two habitable planets in separate orbits around it is small for stars with complete ecospheres. Earlier in the book (see page 155) we said that the probability of two planets of any kind in the ecosphere was 0.036 (1 chance in 28). The probability that both planets fulfill all other requirements and are habitable is only 0.0013 (1 chance in 800). Even this chance decreases rapidly for stars with masses less than 0.88 times that of the Sun, where the width of the ecosphere is progressively narrowed.

The probability of paired habitable planets circling one another about a star is even smaller; and the probability of two sets of paired habitable planets about a star is much smaller still. Such configurations are not impossible. But the maximum possible number of habitable planets around a single star is probably four (arranged in two pairs), and we are extremely unlikely to find such a case quickly.

Now that we have the probability of finding a habitable planet among stars of a given spectral class, we must next determine how many such habitable planets are to be found in the Galaxy. For this we need factor 1. In our discussion of the prevalence of stars of suitable mass (see page 147), we concluded that the stars of spectral classes F2 to K1, inclusive, made up 13 per cent of all stars in the Galaxy. Since it is estimated that there are 135 billion stars in the Galaxy, it follows that there are 17.5 billion stars of suitable spectral class in the Galaxy.

The probability that stars within this range

of spectral classes possess habitable planets varies with the particular spectral class, as shown in the list on page 169; but the average probability for all the spectral classes on the list is 0.037, or about 1 in 27. This means that there are, roughly, 600 million habitable planets in our Galaxy.* There are also, to be sure, countless millions of other galaxies in the Universe (the usual estimates are in excess of 100 billion, in fact), and each is expected to have its own quota of habitable planets.

However, there is little point in considering habitable planets in other galaxies when our own lies all unexplored before us. Indeed, considering that we are only at the threshold of space flight, we are particularly interested in the numbers and possible locations of habitable planets in our own immediate portion of the Galaxy.

To reach a decision on this, let us consider that in our own neighborhood of the Galaxy, it is estimated that there is one star for every 400 cubic light years. This means that there is only one star in the appropriate spectral range (only 13 per cent of all stars) for every 3000 cubic light years. And since only about 3.7 per cent of the stars in the proper spectral range have habitable planets, there should be 1 habitable planet for every 80,000 cubic light years. If a sphere is

* This also means that 1 star out of every 200 in the Galaxy, counting all types and spectral classes, is accompanied by a habitable planet. This is a respectable figure, considering the rather rigid restrictions we have set on habitability out of consideration for human comfort.

drawn about our Solar system with a radius of 27 light years, the volume of that sphere will be 80,000 cubic light years. Within that sphere, then (assuming that habitable planets are spread evenly through our neighborhood of the Galaxy), we might expect to find one habitable planet (other than our own, of course).

If we extend the sphere, we may expect to find 2 habitable planets in a sphere with a radius of 34 light years, 5 in a sphere with a radius of 46 light years, 10 in a sphere with a radius of 58 light years, and 50 in a sphere with a radius of 100 light years. Actually, 100 light years is a small distance considering that the thickness of our Galaxy is over 10,000 light years at the center and that its diameter is about 80,000 light years. To find 50 habitable planets within 100 light years is, therefore, a rather pleasant prospect.

Throughout the Galaxy, the average distance between a given star, chosen at random, and its closest stellar neighbor is about 4 light years; and the average distance between a star with a habitable planet and its closest neighbor with a habitable planet is about 24 light years (following the analysis of probabilities presented in this book). There are regions within the Galaxy, however, where stars are strewn far more thickly than in our neighborhood, and where habitable planets are in far closer reach of one another. In the central regions of the Galactic nucleus and of globular clusters, the average separation between neighboring stars may be in tenths or even hundredths of light years. Therefore, mean separations between habitable planets (if planets can exist in stable orbits in these densely populated regions) could possibly be measured in mere hundreds of billions of miles. If mankind ever penetrated to these regions, its further expansion might then reach explosive proportions.

For the present, however, we must consider only our own corner of the Galaxy and be content with its sparseness. Just where, in this corner, shall we search for habitable planets?

The Nearest
Candidates

In attempting to answer the question with which
the previous chapter closed (involving the where-
abouts of the closest habitable planets), let us
consider the stars that lie within 22 light years
of the Sun. These include 100 stars visible by
naked eye or telescope, plus 11 more that represent unseen companions detected by other than
directly optical means. This makes 111 near
neighbors altogether. Sixty-eight of these 111 can
be omitted at once as having no reasonable likelihood of possessing habitable planets, for the following reasons:

Three (Sirius, Procyon, and Altair) are excessively massive and therefore too short-lived.

Seven are white dwarfs and have gone
through catastrophic stages in development that

surely destroyed any habitable planets that might ever have existed about them.

Fifty-seven are too small and would have too powerful a tidal braking effect on any planet close enough to be at habitable temperatures.

Finally, one (40 Eridani A), otherwise acceptable, is in a system with a nearby white dwarf, a fact that produces inadmissible complications for any habitable planet in the neighborhood.

This leaves some 43 of our nearest neighbors with at least some chance of possessing habitable planets. Of these 43, however, one (Lalande 21185 A) is borderline. It is in a multiple star system of which the orbital characteristics (though not well established) would, if current estimates are correct, be incompatible with the existence of stable planetary orbits within an ecosphere. An additional 28 are borderline in the sense that they are so small that a habitable planet could only exist if it possessed a satellite large enough and close enough to maintain its rotation rate in the habitable range.

If we omit the borderline cases, too, we are left with 14 stars among our immediate neighbors, each with a probability of possessing a habitable planet of over 0.01 (better than 1 chance in 100, in other words).

Let us take up each of these 14 separately, in the order of their increasing distance from the Solar system.

Alpha Centauri A and Alpha Centauri B

Because of its location close to the south celestial pole, the Alpha Centauri system, consisting of 3 stars, cannot be seen from positions on the Earth's surface north of latitude 30° N (roughly the latitude of New Orleans, Louisiana; Cairo, Egypt; and Shanghai, China).

The apparent orbit of Alpha Centauri B around Alpha Centauri A, as obtained from numerous telescopic observations over the past century, is extremely elongated, since it is seen almost edge-on from the Earth. The actual orbit has an eccentricity of 0.52, which means that the distance separating the stars differs considerably with the position of Alpha Centauri B in the orbit. At the point of closest approach (periastron), the 2 stars are separated by about 1 billion miles, somewhat more than the distance of Saturn from the Sun. At the point of farthest separation (apastron), the distance between the 2 stars is about 3.3 billion miles, somewhat more than the distance of Neptune from the Sun.

Even at periastron distance, the separation of the 2 stars is sufficiently large to allow each to have a full ecosphere, undisturbed by the other. Although there is no theoretical method of completely determining the stability of orbits of planetary bodies in multibody systems, approximate methods of treatment suggest that habitable planets (if any) orbiting within the ecospheres of Alpha Centauri A or Alpha Centauri B should have highly stable orbits.

The larger of the two stars, Alpha Centauri A, is a star very similar to the Sun. Its spectral class is given as G4 (or sometimes as Go); its apparent visual magnitude of 0.09 makes it the brightest star (in appearance) of any of the stars on our list of candidates. Its mass is about 1.08 times that of our Sun, and the probability of its possessing a habitable planet is about 0.054 (or about 1 chance in 19).

Alpha Centauri B is somewhat smaller, of spectral class K1 (or K5). It has an apparent visual magnitude of 1.38 and a mass 0.88 times that of our Sun. Its probability of possessing a habitable planet is about 0.057 (or 1 chance in 18).

The third component of the system, Alpha Centauri C (often called Proxima Centauri), is far distant from the other two and can have no possible effect on the habitability of planets circling the larger components. It is itself far too small to have even the faintest chance of possessing a habitable planet.

Since any traveler reaching Alpha Centauri A will also, in effect, have reached Alpha Centauri B, he would reasonably be interested in the probability that a habitable planet would be found circling one or the other (and he could scarcely care which). The probability that one of the two possesses a habitable planet turns out to be 0.107, or just about 1 chance in 9.3. As it happens, this is the highest probability for any star or star system on our list, and it is rather a stroke of luck that this probability is to be found for the star nearest to us. We have our best single chance at the cost of the shortest possible trip.

Epsilon Eridani

Located in the sky about 10° south of the celestial equator, Epsilon Eridani can be seen from any part of the Earth's land surface, except for some far Arctic regions. It is an isolated star (no companion has ever been detected) of spectral class K2 (or K0). Its apparent visual magnitude is 4.2, and its mass is estimated to be about 0.80 times that of the Sun. The probability that it possesses a habitable planet is 0.033 (or 1 chance in 30).

Tau Ceti

Tau Ceti is fairly close to Epsilon Eridani in the night sky. (It is in a neighboring constellation and only slightly farther south.) Like Epsilon Eridani, Tau Ceti is apparently an isolated star. Its spectral class is variously given as G4, G8, and K0. Its apparent visual magnitude is 3.65, and its mass is estimated to be 0.82 times that of the Sun.

The probability that it possesses a habitable planet is 0.036 (1 chance in 28).

Both Epsilon Eridani and Tau Ceti were "listened to" in the spring of 1960 during the course of what was called "Project Ozma," an attempt to detect information-bearing radio signals directed toward our Sun by possible intelligent inhabitants of planets of these stars. An 85-foot-diameter radio-telescope at Green Bank, West Virginia, was used. Radio waves with a wavelength of 21 centimeters were listened for, because this is the wavelength emitted by the neutral hydrogen atoms that make up most of the interstellar matter of the Galaxy. It was reasoned that radio astronomers on any planet would be interested in that wavelength and would assume that any radio astronomers on other planets would be equipped to receive it.

No signals were detected. In a way, this is not surprising. The joint probability that Epsilon Eridani and Tau Ceti might have one habitable planet between them is estimated to be about 0.07 (or 1 chance in 14). The probability that a given habitable planet will be inhabited by intelligent beings at a sufficiently advanced stage of technology to beam strong radio signals out into space is difficult to estimate, but certainly it must be quite low.

It should be remembered, after all, that if the reverse experiment were conducted, it would be no more successful. If creatures on some planet orbiting around Epsilon Eridani or Tau Ceti had directed their radio telescopes toward our Sun, they would have heard no radio signals betokening life. We are not sending out signals on a wavelength of 21 centimeters for them to detect. Indeed, until the last half-century of the hundred thousand or more years during which highly intelligent life has existed on the Earth, such radio waves would have been beyond our capacity to send. For all of these reasons, the negative results of Project Ozma represent no cause for pessimism and give us no reason to cut down any of the probability estimates in this book.

70 Ophiuchi A

The system of 70 Ophiuchi consists of two stars revolving around each other with a period of 87.85 years in an orbit with an eccentricity of 0.50. The orbits described by the two stars are almost identical with those described by Alpha Centauri A and Alpha Centauri B. No third companion has been established for the system, although dark companions are suspected. The brighter star, 70 Ophiuchi A, has an apparent magnitude of 4.19 and a mass about 0.90 times that of our Sun; hence, it has a 0.057 probability of possessing a habitable planet (or 1 chance in 18).

The less massive component, 70 Ophiuchi B, is of spectral class K5, with a mass about 0.65 times that of the Sun. It is a borderline case, capable of supporting only a satellite-stabilized planet. (As seen from the system of 70 Ophiuchi, the Sun would appear as a third-magnitude star in the constellation Orion, not far from the belt.)

Eta Cassiopeiae A

The binary system of Eta Cassiopeiae has a period of the order of 500 years and an orbital eccentricity of 0.53. (The existence of a third component is not well established.) The orbits of the two stars would have the same relative shapes as those of the two stars of the Alpha Centauri and 70 Ophiuchi systems, but the stars of the Eta Cassiopeiae system would be separated by some 3 times the distance.

Eta Cassiopeiae A has an apparent magnitude of 3.54, a spectral class F9, and a mass about 0.94 times that of our Sun. The probability of its possessing a habitable planet is 0.057 (1 chance in 18). The smaller component, Eta Cassiopeiae B, is of spectral class K6 and has a mass of 0.58 times that of our Sun. Like 70 Ophiuchi B, it is a borderline case, with very little chance of possessing a habitable planet. (Our Sun, as seen from the Eta Cassiopeiae system, would appear to be imbedded in the Southern Cross.)

Sigma Draconis

This is the most northerly star in the list and appears to be an isolated star. Its apparent magnitude is 4.72, and it is in spectral class G9, with a mass about 0.82 times that of our Sun. There is a probability of about 0.036 (1 chance in 28) that it has a habitable planet.

36 Ophiuchi A and 36 Ophiuchi B

This system, consisting of 3 stars, lies almost directly between us and the center of our Galaxy. The orbital elements of the system have not yet been established. The brightest star, 36 Ophiuchi A, is in spectral class K2, has an apparent magnitude of 5.17, and a mass about 0.77 times that of the Sun. Thus it has a probability of about 0.023 (or 1 chance in 43) of having a habitable planet. As for 36 Ophiuchi B, it is in spectral class K1, with a mass about 0.76 times that of the Sun. Its probability of having a habitable planet is 0.020 (or 1 chance in 50). This pair of stars is very close indeed to being an example of "twin suns."

Here, as in the Alpha Centauri system, we might calculate the chance of a habitable planet orbiting around either of the two brighter components of the system. This probability comes to 0.042 (1 chance in 24).

Finally, the third component, 36 Ophiuchi C, in spectral class K6, has a mass about 0.63 times that of our Sun and is another borderline case for which the probability of a habitable planet is very small.

HR 7703 A

This system is in the southern constellation Sagittarius, and it consists of two stars for which the orbital elements have not yet been determined. The brighter star, HR 7703 A, has an apparent magnitude of 5.24, is of spectral class K2, and has an estimated mass 0.76 times that of the Sun. The probability of its having a habitable planet is 0.020 (1 chance in 50). The smaller component, HR 7703 B, is of spectral class M5 and is far too small to possess a habitable planet.

Delta Pavonis

Even more southerly than Alpha Centauri, Delta Pavonis cannot be seen by observers on the Earth's surface north of latitude 23° N (that is, from anywhere in the North Temperate or North Frigid Zones). It is apparently an isolated star of spectral class G7. Its apparent magnitude is 3.67, and its mass is 0.98 times that of our Sun. The probability of its possessing a habitable planet is 0.057 (or 1 chance in 18).

82 Eridani

Another apparently isolated star, 82 Eridani, cannot be seen from latitudes north of 46° N (the latitude of Portland, Oregon). Its spectral class is G5, its apparent magnitude is 4.3, and its estimated mass is 0.91 times that of our Sun. Its probability of possessing a habitable planet is 0.057 (or 1 chance in 18).

Beta Hydri

This is the most southerly star on our list of candidates, and cannot be seen north of latitude 10° N (the latitude of the Panama Canal). It is an isolated G1 star with an apparent visual magnitude of 2.90 and an estimated mass 1.23 times that of our Sun. It has a probability of 0.037 (1 chance in 28) of possessing a habitable planet.

HR 8832

This is the faintest and most distant star on our list. It is an isolated star in the constellation Cassiopeia. It is in spectral class K3, has an apparent magnitude of 5.67, and an estimated mass 0.74 times that of the Sun. The probability of its possessing a habitable planet is only 0.011 (1 chance in 90).

Over-all View of the Sun's Neighborhood

To summarize the 14 stars on our list, we present them now, in order of distance:

Star	Distance from the Earth (in light years)	Probability of a Habitable Planet
Alpha Centauri A	4.3	0.054 ⎫ 0.107
Alpha Centauri B	4.3	0.057 ⎭
Epsilon Eridani	10.8	0.033
Tau Ceti	12.2	0.036
70 Ophiuchi A	17.3	0.057
Eta Cassiopeiae A	18.0	0.057
Sigma Draconis	18.2	0.036
36 Ophiuchi A	18.2	0.023 ⎫ 0.042
36 Ophiuchi B	18.2	0.020 ⎭
HR 7703 A	18.6	0.020
Delta Pavonis	19.2	0.057
82 Eridani	20.9	0.057
Beta Hydri	21.3	0.037
HR 8832	21.4	0.011

(Naturally, the above estimates of probabilities are very uncertain and are subject to future revision.)

The combined probability of the existence of at least one habitable planet in the whole volume of space out to a distance of 22 light years from the Sun is about 0.43. In gambler's parlance, we might say that the odds were about 3 to 2 against our finding even a single habitable planet in the entire list of 14 candidates presented above (to say nothing of the remaining 97 stars, seen and unseen, that occupy the volume of space within 22 light years of the Sun).

Still, those are not such bad odds, considering that we are dealing with but a trifling corner of the Galaxy—our own doorstep, so to speak. Once we learn to make our way among the stars, we will undoubtedly be prepared to go much farther than 22 light years, and out in the rest of the Galaxy, 600 million habitable planets (by our estimate) await us.

/

Star Hopping

Detecting a Planet

We may be able to calculate, from our arm-chairs, the probability of a given star possessing a habitable planet in orbit around it, but under what conditions would we be certain of it, one way or the other? Naturally, a space flight to the star in question would tell us, but need we actually travel all the way there?

In other words, at what distance can it be ascertained that a specific star actually possesses an apparently habitable planet? And how closely must we approach a planet in order to be sure that it is indeed habitable?

As we know from experience, it is not easy to tell a great deal about a planet even from a vantage point within the stellar system of which it forms a part. There are still many unanswered questions, for instance, about Venus and Mars, although the Earth periodically comes closer to them than 30 and 40 million miles, respectively. To be sure, this is not grounds for complete pessimism. Much of the difficulty with respect to our neighbor planets lies in the fact that we are looking at them through our own radiation-absorbing and image-distorting atmosphere. Undoubtedly, we will learn a great deal more about the planets of our own system (even before we actually visit them) once telescopes of merely moderate size are put into operation in space above the atmosphere.

Imagine, then, that we are on an exploratory mission on a spaceship that is approaching a star. We have telescopes up to 60 inches in diameter, sensitive radio receivers, and other necessary equipment. We would like to make a decision at the earliest possible moment whether to continue going toward the star or whether to break off and return to base, or head for a second objective.

Let us say that our first decision will be based on whether the star has a planet orbiting within an ecosphere. If it has such a planet, we will proceed; if we can say that it definitely does not, we will break off. To see what this implies, let us consider a star like our Sun about which planets like Jupiter and the Earth circle at distances equal to those that hold true in our Solar system. At what distance could we detect the Jupiter-like planet, and at what distance the Earth-like planet?

In a way, we can detect Jupiter-like planets even at distances of a number of light years. In recent years, we have actually detected three planets of this sort without moving from the surface of the Earth. In 1948, the proper motion of the binary 61 Cygni system showed irregularities that could be due only to the presence of a third non-luminous object, 61 Cygni C. From the size of the irregularities of motion, it was deduced that 61 Cygni C had a mass only 0.008 times that of the Sun and, therefore, 8 times that of Jupiter.

This new body, 61 Cygni C, is larger than any planet in our Solar system; in fact, it is large enough to be in the planet-to-star transition range (see page 50), which alone would make it a fascinating object for close-range study for anyone interested in general planetology. If we assume that it is a giant planet rather than a tiny star, then it has the distinction of being the first planetary body to be discovered outside our Solar system.

In 1960, similar studies showed that Lalande 21185 A had an invisible companion, Lalande 21185 B. The latter body, like 61 Cygni C, is about 8 times as massive as Jupiter. Finally, in 1963, a smaller body, one that can only be a planet, was found to be circling Barnard's Star. This new discovery, Barnard's Star B, is estimated to be only 1.5 times the mass of Jupiter.

Barnard's Star A is only 6 light years from us, and Lalande 21185 A is only 8 light years distant. They are the second- and third-closest star systems, respectively, and only the Alpha Centauri system is closer. The fact that two out of three of the nearest star systems have already been found to possess planetary bodies (despite the handicap of observations from great distances) is strong evidence for the contemporary feeling that all stars have non-luminous companions. To be sure, the planets discovered cannot possibly be habitable, and indeed none of the stars involved could have habitable planets (except for possible satellite-maintained planets around either 61 Cygni A or 61 Cygni B); but what is true for these stars is also true for stars better suited for habitable planets.

The detection of planets by means of their gravitational effects upon their primaries, while the only route open to us now, would be impractical for our exploring space travelers. Discovery through gravitational effects requires many years of meticulous observation. We must resolve the problem more quickly than that, and the most practical way of doing so is actually to see the planet.

When viewing the Universe through our atmosphere, it is very difficult to observe a faint object close to a much brighter object. This is because the light from the brighter object interferes and because the poor "seeing" through our shifting, wavering atmosphere prevents us from obtaining the best results that, in theory, our large telescopes could give us. A telescope employed in space, however, is free of atmospheric interference and would be able to attain its theoretical best in detecting and separating closely spaced objects of equal brightness. Also, faint objects close to brighter ones might be picked up through use of special techniques such as hiding the image of the brighter object with a knife edge.

It seems that such techniques could be used to reveal a dim body within 2 seconds of arc* of a bright one—provided the apparent magnitude of the dim body was +18 or brighter, since a 60-inch telescope cannot detect anything dimmer than a magnitude of +18.

If we assume that a planet like Jupiter is viewed by a spaceship that is approaching from an angle that makes it possible to view the planet as widely separated from its primary as possible, the planet would then be half illuminated to the eye. The maximum separation of the planet and its primary (assuming the separation to be that between Jupiter and the Sun) would be 2 seconds of arc at a distance of 8.55 light years. This is twice the distance that separates us from the Alpha Centauri system.

However, a Jupiter-like planet could be seen

---

* An arc of 2 seconds is slightly less than the apparent diameter of a United States dime when viewed from a distance of one statute mile.

at that distance only if it had attained the necessary minimum brightness, and this it certainly
does not manage at a distance of 8.55 light years.
Assuming an albedo* of 0.4 and an image that is
half illuminated, its magnitude would not reach
the necessary +18 until a distance of 0.43 light
years is reached. This is a distance of 2.5 trillion
miles, or 600 times the distance of Pluto from the
Sun.

Thus, it is the apparent magnitude and not
the apparent separation from its primary that sets
the maximum distance for optical detection. At
a distance of 0.43 light years, the separation of
planet and primary would be about 37 seconds
of arc.

Seeing the planet as a mere speck of light
would still not be enough, however. It would be
imbedded in a field or background containing so
many distant stars of like magnitude that each
such object would have to be analyzed spectroscopically to discriminate between the planet and
the surrounding stars; or else repeated photo-
graphs would have to be taken to detect the motion of the planet against the background of
motionless stars. Much, too, would depend on the
tilt of the system with respect to the line of approach. The most favorable approach line would
be at right angles to the plane of the planetary
orbit, for then the planet would always be seen

---

* The albedo is the fraction of the incoming light that is reflected by a planetary body.

at maximum distance from its primary. Under less favorable conditions of approach, with the planet more or less between us and the star, the apparent distance between the two would be considerably less than maximum much of the time, and detection could be made only at distances much shorter than 0.43 light years.

All in all, in case of continuing doubt, it might be necessary to come in close enough to see the Jupiter-like planet as a disk. Such a disk would be discernible in a 60-inch telescope as an object 2 seconds of arc in diameter at about 90 A.U. (8.4 billion miles, or somewhat more than twice the distance of Pluto from the Sun).

Similar difficulties, of a more extreme type, would prevail in the detection of a habitable Earth-like planet, considerably smaller than Jupiter and considerably closer to its primary. Assuming an Earth-like albedo of 0.36, and maximum separation, such a planet would be separated from its primary by an apparent distance of 2 seconds of arc at a distance of 1.7 light years, or about a third of the way to the Alpha Centauri system. However, the Earth's reflected light would be detectable only at one-tenth that distance, 0.17 light years or just about 1 trillion miles. An Earth-like body would be perceptible as an actual disk of 2 seconds of arc in diameter at a distance of about 7.6 A.U. (700 million miles), or not quite the distance between Saturn and the Sun.

We would not, then, be able to detect the

presence of an Earth-like planet (under the present assumptions) until we had approached the target system to within one-sixth of a light year. Unfavorable indications, such as the presence of a giant planet in such a position as to prevent the orbiting of a habitable planet within an ecosphere, might be detected earlier.

Detecting Planetary Details

Once an approach to within one-sixth of a light year has been made and an Earth-like, *possibly* habitable planet has been detected, one must approach even more closely to *ascertain* habitability. (If no such planet is detected, of course, after so close an approach, it may pay the explorers to veer away.)

As one continues the approach, various properties will be observed that will contribute to a decision concerning its habitability beyond that of the mere existence of a planet of the proper size within an ecosphere. Radio waves, for instance, might be detected of a form that would bespeak not only a habitable planet, but one with intelligent life.

Radio waves in the "broadcast band" (535 to 1605 kilocycles), originating on Earth, do not penetrate the Earth's ionosphere very effectively, and most of the energy is refracted and returned to the Earth. It is only the very-high-frequency radio waves (those with frequencies of more than

30,000 kilocycles, or 30 megacycles) that would be detectable in space. In the very-high-frequency bands, some of the most powerful Earthly sources are the BMEWS (Ballistic Missile Early Warning System) radar stations broadcasting signals of a frequency of about 400 megacycles. Taking the peak power level as about 1 megawatt (1 million watts) and assuming that the natural Galactic noise at the receiver is that produced by objects at room temperature, the signals from BMEWS transmitters might be detected at a distance of 10 billion miles (or 2.5 times the distance of Pluto from the Sun).

Such a situation would not be very likely, however, since we do not expect to find habitable planets with intelligent life (as opposed, simply, to life) except at rare intervals. Still, planets may emit radio signals even in the absence of intelligence. Certain kinds of natural radio noise may be characteristic of planets that, like the Earth, have numerous thunderstorms and lightning flashes occurring in their atmospheres at all times. There might be a recognizable pattern to such radio noise that would at least serve as an indicator of a planet with an atmosphere, though not necessarily a breathable one, of course. What fraction of this noise would leak out and be detectable in space at a distance from the source is not known with any degree of certainty.

Characteristic spectral absorption lines due to oxygen or water vapor might be detected as soon as a planetary image is obtained. The detection of these lines can be accomplished only with great difficulty when looking through the Earth's atmosphere because the prevalence of oxygen, water vapor, and other absorbing gases in our atmosphere confuses the issue; but such detection would be quite feasible from space.

Estimates of the approximate distances at which important surface features of the Earth could be identified from space are oceans at about 7 A.U. (650 million miles), forests at about 3 A.U. (280 million miles), and large cities at about 5 million miles, under good viewing conditions. Closer approaches and very clear atmospheres would be needed in order to identify as artificial structures such conspicuous works of man as canals, dams, bridges, airfields, railroads, and so on.

Generally speaking, then, we will not be able to explore the stars cheaply by taking a quick look from a distance. A system will have to be approached to within one-sixth of a light year before it can be definitely determined to have no habitable planet (although certain negative indications might be obtained while the exploring party is still as far away as one-third to one-half of a light year). This means that even the nearest star system cannot be eliminated until 90 to 95 per cent of the trip there has been completed. For more distant stars, the portion of the trip that must be completed is higher still.

Once the presence of a habitable planet is seen to be possible, a closer approach to within a million miles or so is required in order to be fairly certain of habitability. To be absolutely certain necessitates a landing on its surface and direct investigation of such matters as its atmosphere and surface conditions.

Any indication that a planet is already inhabited by intelligent creatures would signal the need for proceeding with the utmost caution. In fact, before a manned landing is made on any likely looking planet, it would be desirable to study the planet thoroughly from a distant orbit around it for a protracted period of time; to send sampling probes into its atmosphere; and to send surveillance instruments down to the surface. Contacts with alien intelligence should be made most circumspectly, not only as insurance against unknown factors, but also to avoid any disruptive effects on the local population produced by encountering a vastly different cultural system. After prolonged study of the situation, a decision would have to be made whether to make overt contact or to depart without giving the inhabitants any evidence of the visitation.

Interstellar Flight

As man's ability to accelerate payloads to higher and higher velocities increases, a point may eventually be reached when interstellar flight can begin to be considered feasible and when the type of exploration that we have been assuming in this chapter will really enter the realm of the possible. At the present time, we have barely reached the point in technology where we could design a rocket vehicle capable of sending a small package completely out of the Solar system. Its velocity at a distance equivalent to the orbit of Pluto would be very small relative to the speed of light, however, perhaps not more than 1/10,000 that speed (18.6 miles per second, or 67,000 miles per hour). Consequently, trips to even the closest stars would take many thousands of years, and it would not be worth while even to consider such a slow trip. But if one has confidence in man's ability to learn—a confidence justified by looking back at the accelerating pace of technical progress over the past 400 years—then one may be optimistic about the future feasibility of flights over interstellar distances.

Flights at velocities that are large fractions of the velocity of light do not violate any of the known laws of physics, although the expenditures of energy needed for accelerating to such velocities are enormous. It is to be hoped, despite that, that flights at such high velocities will some day become practical.

Once we have a capability for sending small packages over interstellar distances at velocities of the order of one-tenth the speed of light or higher, then we could consider sending unmanned probes to the vicinity of the nearest promising candidate stars to report back to Earth on the prevailing conditions. (It would take years for radio waves, or any other form of communication, to make their way back to the Earth, but that cannot be helped.) Favorable reports could be followed by manned expeditions to those systems showing the most promise in the light of our then greatly enhanced knowledge of astrophysics and general planetology.

The requirements for speed and ingenuity will depend on how far away the nearest promising star-candidate happens to be and on how determined people are to make the trip. If one is willing to spend 20 years en route, then trips to stars 4 light years away could be made at one-fifth the speed of light (37,000 miles per second); trips to stars 10 light years away could be made at one-half the speed of light (93,000 miles per second); trips to stars 15 light years away could be made at three-quarters the speed of light (140,000 miles per second); and so on.

At velocities approaching the speed of light, moreover, relativistic time-contractions become quite evident, so that for a particular round trip considerably less time would seem to have elapsed for the traveler himself than for an Earthbound observer. Thus, enormous distances (from the viewpoint of the Earthbound observer) could be traversed in 20 years (from the standpoint of the traveler) if velocities very close to the speed of light are attainable.

There is, it should be pointed out, a complication in the optical detection of objects in space when the observer is moving at near-light velocities. At such velocities, the light from the star being approached seems to shorten considerably in wavelength and becomes quite a bit more blue. In other words, the star appears to be hotter than it actually is. On the other hand, when one is moving away from a star at such velocities, its light seems to shift toward the red end of the spectrum, and the star appears to be cooler than it is.

At very high velocities, a black spot will appear in the directly forward direction, since all the visually detectable radiation from the stars in that part of the field of view would be shifted into the ultraviolet. Another black spot will appear in the directly backward direction, since all the visually detectable radiation from the stars in that sector would be shifted into the infrared. Between the two black spots, the stars will range in color from blue to red as one progresses forward to backward. In the visual black spots, however, the stars will still be detectable with instruments.

These effects would start to become apparent to the casual observer at velocities of about one-twentieth the velocity of light (9300 miles per second), while the black spots would appear at

velocities of about one-half the speed of light
(93,000 miles per second).

Even at vehicle velocities much lower than
that of light, however, the duration of the trip
might possibly be shortened (for the traveler)
through the development of hibernation techniques or their equivalent. The traveler might
then sleep the years away as his ship approached
its destination.

Kinds of Habitable Planets

Once interstellar flight is developed and stars
with habitable planets are reached, unlooked-for
varieties of experience should open to the human
race.

To be sure, the most common kind of habitable planet, if the ideas developed in this book are
essentially correct, should be similar to the Earth
in many respects. (The one mild peculiarity of
the Earth is its rather large natural satellite.)
The typical habitable planet should be of much
the same mass as the Earth, although perhaps a
bit smaller on the average, and it should have
a similar atmosphere, a similar night-day cycle, a
sun of similar size and appearance, a mild inclination to its equator, and a moderate eccentricity to
its orbit.

Seasons should be part of the common experience of the inhabitants, as well as oceans, beaches,

winds, clouds, rain, snow, thunder, lightning, volcanoes, earthquakes, rainbows, sunsets, starry
nights, blue skies, deserts, mountains, rivers, gla-
ciers, and polar ice-caps. In short, most of the
physical and meteorological phenomena with
which we are familiar should also be known on
most other habitable planets, although always
with enough difference in detail to make each
planet a fascinating experience.

When it comes to the living things indigenous
to the planets, of course, these might differ widely,
depending on the precise course that evolution
happened to take in each special circumstance.
Even so, on each planet one would expect to find
organisms carrying on photosynthesis and creatures capable of invading practically every conceivable corner of the environment: marine forms,
fresh-water creatures, land creatures, aerial forms,
cave organisms, and so on. In spite of differences
in detail, certain basic kinds of life forms would be
expected to display some common characteristics.
Thus, fast-swimming marine forms would be
streamlined, land animals would typically have
legs, and fast-moving aerial forms would have
wings. There must of necessity be autotrophs (life
forms that use only inorganic nutrients), and one
would also expect to find heterotrophs (life forms
that use autotrophs or other heterotrophs for
food).

On no other planet, however, would we expect to find any of the actual phyla of plants or

animals with which we are familiar on the surface
of the Earth. From the smallest virus to the largest
whale, the Earthly life forms are unique products
of the Earth. Each planet on which living things
have evolved must have its own peculiar classification of organisms, and this in itself should present
the human race in general and biologists in particular with endless realms of new wonder and experience.

But in addition to these typical Earth-like
habitable planets, there should also be many unusual and rarer kinds among the 600 million habitable planets that we have estimated to exist in our
Galaxy. Some of the special types are easily imagined.

A *satellite planet* would be a habitable planet
in orbit around a gas giant like Jupiter, the two
revolving about a primary. The rotation of the
habitable planet would be stopped with respect
to the companion but rotating with respect to the
primary. Such a planet would have unusual cycles
of light and dark on the side facing its companion,
for the nights would be dominated by the presence
of an extremely large and luminous "moon." (The
side away from the companion would have more
normal day-night cycles.) Then, too, eclipses of
the primary would occur every day unless the
planet's orbit about its companion was at a marked
angle to that of the companion's orbit about the
primary.

Twin planets, revolving about a common cen-

ter of mass, would have their rotation stopped
with respect to each other. It would be interesting
to speculate on the intellectual development of a
life form on one planet of such a pair. With another planet, close and large, with clouds, oceans,
and continents clear to the naked eye, could any
egocentric philosophy of the Universe develop?
Would the urge to reach the companion hasten
technological progress? We can ask but can offer
no answer.

A *planet with two suns* would be a habitable
planet in orbit around two stars that are very close
to each other. Close binaries (separated by a few
million miles, say) would produce a somewhat
complicated sunrise and sunset pattern for a habitable planet in orbit around them and an interesting variation in light intensity as the stars eclipsed
each other; but otherwise they would not necessarily greatly affect conditions on the surface of
the planet.

A *planet in a widely separated binary system*, rotating about one star in that system, would
have very bright nights during those parts of the
year when it was passing between the two stars,
and "normal" dark nights only when the companion, as well as the primary, was beneath the
horizon. From a habitable planet in orbit around
Alpha Centauri A, for example, Alpha Centauri B
would seem to be a very bright star-like object
that would range in magnitude from -18 to -20,
depending on whether the two stars were at

apastron or periastron. Alpha Centauri B, at its brightest, would appear over 2 million times as bright as Venus appears to us and nearly 1500 times as bright as our full Moon. (This type of planet and the type mentioned immediately preceding may not be particularly rare.)

A *planet with very high equatorial inclination* must lie within a narrow range of distances from the primary to be habitable at all, and even then only a small fraction of its surface might be habitable. A planet with an equatorial inclination of 75°, even at optimal distance from its primary, would be habitable only between latitudes 14° N and 14° S. On Earth, this would represent a belt 2000 miles wide, centered on the equator. Higher latitudes would be excessively cold during the winter season and excessively hot in the summer season.

A *planet with two habitable belts* would most likely be found among planets having low equatorial inclinations but orbiting near the inner edge of the ecosphere. Such planets are excessively hot near the equator; hence the habitable regions would be in the intermediate or high latitudes only. A planet having the same equatorial inclination as the Earth (23.5°) but orbiting at a distance from its primary such that it received 30 per cent more illuminance than the Earth would be habitable only between latitudes 51° to 66° N and S. (Such belts on the Earth would include the Soviet Union, Canada, the British Isles, Scandinavia, and parts of Germany and Poland in the

north, and nothing in the south but the extreme
tip of South America.) With the wide belt from
51° N to 51° S intolerably hot and perhaps im-
passable to many life forms, it is quite probable
that land life forms would evolve more or less independently in the two regions. Marine life and
some aerial forms might migrate between the two,
but land migrations would be fairly effectively
stopped by a heat barrier consisting mainly of
deserts, possibly with small pockets of habitable
territory on mountain plateaus.

   Habitable planets with rings are another possibility. There are many unknowns associated with
the origin and composition of the beautiful ring
system of Saturn, and it is quite probable that
massive oblate planets are more likely to possess
rings than are planets in the habitable class. Nevertheless, it is reasonable to suppose that a few
habitable planets may also have flat equatorial
rings inside their Roche limits, although these
rings would probably not be as thick with debris
as are the rings of Saturn.

   Other special types of habitable planets may
also exist. Since oceans are believed to be products
of volcanic activity, and since such activity increases, possibly, with the mass of a planet, it may
be that high-g planets are largely oceanic and
low-g planets are largely dry land. Planets with
more extensive oceans than the Earth—those having, say, 90 per cent oceans and 10 per cent dry

land—might well have quite widely separated continents with no land bridges. Fairly independent evolutionary courses might be followed by the land life forms, so that each continent would be virtually a world to itself, biologically speaking. On the other hand, planets having substantially less oceanic water than the Earth might well have non-interconnecting oceans (large lakes, really) with isolated forms of marine life undergoing separate evolutions.

An oceanic planet with the same irradiance as the Earth would have more equable temperatures, so there would be an absence of polar ice-caps. A dry-land planet, in the absence of worldwide oceanic circulation, would have extreme temperatures. A high fraction of the land surface probably would be desert, and the main habitable regions would tend to be close to the landlocked seas.

Our Sun is situated in a fairly sparsely populated section of the Galaxy—in one of the spiral arms well out toward the edge of the disk. Our night sky, therefore, is relatively lightly sprinkled with stars in comparison with what would be seen from planets located in certain other positions in the Galaxy. Something like 2500 stars brighter than magnitude 6.5 are visible to the naked eye from a given spot on the Earth on a clear night. Much more spectacular night skies would be seen from habitable planets located in globular clusters or near the center of the Galaxy. It has been estimated that about 2 million stars of magnitudes brighter than 6.5 would exist above the horizon of a habitable planet located near the Galactic center. The starlight in such a sky would be roughly equal to the light of the full Moon as seen from the Earth. The scattered light in such a sky would, however, prevent one from seeing any but the brightest stars. Stars fainter than, perhaps, magnitude 2.5 could not be seen at all against the over-all luminosity of the night sky. Even so, the number of stars brighter than this limiting magnitude would number approximately 30,000, or more than 10 times as many as we can see on the darkest night.

On the other hand, habitable planets around stars imbedded in some of the dusty regions (dark nebulae) of the Galaxy might have almost no stars at all in their night skies. And habitable planets around stars at the very edge of the Galaxy would have stars in one half of the celestial sphere but none in the other half. On the starry side, the Galaxy as a whole would be visible as a Milky Way with structure (a foreshortened vision of a spiral Galactic whirlpool). The only lights in the night sky looking away from the Galaxy (apart from local planets and a very occasional star) would come from extra-galactic nebulae or distant island universes, only a few of which, like the Large and Small Magellanic Clouds, or the Great Nebula in Andromeda, would be faintly visible to the naked eye.

Man in New Environments

And if new habitable planets offer new kinds of environments for man, man may well respond by developing new varieties of himself.

Human beings, together with all other life forms on the Earth, are very well adapted to what we think of as a normal environment. Although the normal environments of the Eskimos, of the Australian aborigines, of the pygmies of Africa, and of the Indians of the high Andes are quite different from one another, they all fall into a fairly narrow range when viewed from the standpoint of possible variations among the different planets. All of the people mentioned above occupy more or less marginal ecological niches, but they have become adjusted to them gradually. This adjustment probably has come about through many generations of selection for individuals who could tolerate exceptionally well the environmental extremes of temperatures, the dietary limitations, the dryness, or the low oxygen partial pressure, as the case happened to be.

Colonizers of other planets will encounter even more varied environmental conditions than those existing on the Earth's surface. The transitions, however, unlike those made by peoples in the past history of the Earth, will of necessity be abrupt rather than gradual. The colonizers will, of course, have the advantages of a very high level of technology to assist them in making the transitions. Even so, unless close contact is maintained with the population pool of the Earth, very profound genetic shifts will undoubtedly occur in a relatively short time, if the environmental conditions are markedly different from those of the Earth.

In future times, if interstellar trips become possible, there may be circumstances under which an expedition locates a habitable planet and then, through accident or design, is cut off from communication with the rest of humanity for several hundred years.

Imagine such a colony marooned on a 1.5-g planet, for instance. Assuming that the colony is able to survive and multiply, there would inevitably be a premium on muscular strength, short reaction times, and accurate judgment in moving about. There would also be a premium on strong internal constitutions.

All this follows from the fact that because of the high surface gravity, so many aspects of life would become more crucial. Accidental falls would be more dangerous, more likely to be fatal or crippling. Sprains, strained muscles, hemorrhoids, fallen internal organs, back, foot, and leg ailments, varicose veins, and certain difficulties of pregnancy would all be more prevalent than in the gentler 1-g environment of the Earth. Thus there would be an inexorable selection pressure continually favoring those individuals best equipped to deal with the problems of a high gravitational force.

After a number of generations under these environmental conditions, the population would change in appearance. Experiments with chickens have shown that animals raised under high gravity exhibit an increase in relative heart and leg sizes, and these observations might also apply to human beings.

Those humans best able to thrive under high-g conditions would probably tend to have shorter arms and legs. They would be more compactly constructed and would have heavier bones than any population normally found on the Earth. Because of the constant drag of gravity, human beings on such a planet would tend to have better muscular development and less unsupported external fatty tissue, so that standards of attractiveness would change. Since it would be an advantage during pregnancy to have small babies, rather than large ones, the average height of the adult population would probably tend to decrease to some optimal level. Again, since objects fall much faster when surface gravity is high, selection would also tend to favor people with unusually rapid muscular reactions.

If isolation is continued for a long enough period of time, there would inevitably be various kinds of minor genetic drifts in unpredictable directions. If continued, enough genetic changes would accumulate so that the isolated population would no longer be able to interbreed with the population of the Earth when the members of the two planets again made contact. In short, new human species could result from interstellar travel. While this would probably require separations of many thousands of years, it would almost certainly occur much more rapidly than it would under conditions of isolation on the Earth.

Other environmental conditions would result in quite different changes to human populations. A colony isolated on a low-gravity planet—one with a surface gravity of, say, 0.75 g—would be exposed to a reduced stress from gravity, but it might also be exposed to a reduced pressure of oxygen in the atmosphere. Selection processes might therefore favor those individuals having more efficient respiratory systems and, possibly, those having larger rib cages, lungs, and respiratory intake. Under less stringent gravitational stresses, those with slender physiques would have no great disadvantage, and any changes in the composition of the population with respect to body type would probably depend on other factors.

On a small planet having a thinner atmosphere and probably a weaker magnetic field than the Earth, the normal background radiation level might well be substantially higher than that at the Earth's sea level, for two reasons. First, as a result of less intense gravitational fractionation of rocky materials in the body of the planet during the formation period, the proportion of heavy minerals, including radioactive materials, in the crust might be higher. Second, with less shielding from cosmic-ray particles from outer space (because of the thinner atmosphere and weaker magnetic field), there would be a greater influx of energetic particles from outside. Accordingly, mutation rates would be expected to be higher, and possibly evolution would proceed at an accelerated rate.

Even without genetic or muscular changes, human capabilities for physical action would be markedly changed on planets having gravitational accelerations different from the Earth's. This may be illustrated with reference to track and field records and various athletic events. For events such as the shot-put, the javelin throw, and the discus throw, the maximum mark attained would vary inversely with the g level. If a champion athlete can throw a javelin 281 feet on the Earth, for example, he could throw it about 375 feet on a planet with 0.75 g and only 187 feet on a 1.5-g planet.

For the high jump, the relationship is not as simple because though a man must lift his own center of mass about a foot above the recorded height of the bar, yet his center of mass is already about 3 feet above the ground when he starts his spring into the air. Taking this into account, we can calculate that if 5.5 feet is the standing high-jump record on the Earth, it would be 7.1 feet on a 0.75-g planet and 3.8 feet on a 1.5-g planet. For running events, the relationships are still less clear, although it may be seen intuitively that men could run faster on low-g planets and more slowly on high-g planets than they can on the Earth's surface.

Finally, the development of modified species of humanity will inevitably broaden the concept of a habitable planet. A human species modified by and acclimated to a 1.5-g planet will, in estimating the habitability of other planets, feel it possible to endure a 1.75-g planet, which we could not. The human species modified by and acclimated to a 0.75-g planet and its thin atmosphere would be able to stretch a point toward further loss of g where we could not. The limits cannot be stretched forever, of course, for it is not likely (for instance) that by gradual acclimatization to lighter and lighter gravities, human beings could ever adapt themselves to a planet so small as to have no atmosphere. Nevertheless, the Galaxy might well, in the end, be inhabited by varieties of men who are not only of separate species but whose criteria of habitability in planets may not be the same.

CHAPTER

8

/ An Appreciation

of the Earth

For all the pictures of far travel and of strange worlds beyond the sky that we have been presenting in this book, it is still on the Earth that we live at present. Surely, however, the vision of the lands beyond will help us look at the Earth with new eyes and with new appreciation.

We take our home for granted most of the time. We complain about the weather, ignore the splendor of our sunsets, the scenery, and the natural beauties of the lands and seas around us. We cease to be impressed by the enormous diversity of living species that the Earth supports. This is natural, of course, since we know only the Earth, and all of it seems very commonplace and normal. We might feel less indifferent, though, if we considered how altered our world would be if some of the astronomical factors were changed even slightly.

Suppose that with everything else being the same, the Earth had started out with twice its present mass, giving a surface gravity 1.38 times Earth normal. Would the progression of animal life from sea to land have been as rapid as it was? While the evolution of marine life would not have differed greatly, land forms would have to be more sturdily constructed, with a lower center of mass. Trees would tend to be shorter and to have strongly buttressed trunks. Land animals would tend to develop heavier leg bones and sturdier musculatures. The development of flying forms would certainly have been different to conform with the denser air (which would produce more aerodynamic drag at a given velocity) and the higher gravity (which would require more lifting surface to support a given mass).

A number of opposing forces would have changed the face of the land. Mountain-forming activity might be increased, but mountains could not thrust so high and still have the structural strength to support their own weight. In addition, raindrop and stream erosion would be magnified, so that the mountains would wear down more rapidly. The steeper density gradient in the atmosphere would change the weather patterns; wave heights in the oceans would be lower, and the reach of ocean spray would be shortened, resulting in less evaporation and a drier atmosphere.

Cloud decks would tend to be lower, and the land-sea ratio would probably be small because the average continental height would be lower, and more water would be produced by volcanic action.

The length of the month would shorten by about 4 days if the Moon's distance remained the same. There would be differences in the Earth's magnetic field, the thickness of its crust, the size of its core, the distribution of mineral deposits in the crust, the level of radioactivity in the rocks, and the size of the ice-caps on islands in the polar regions.

Conversely, suppose that the Earth had started out with half its present mass, resulting in a surface gravity 0.73 times Earth normal. Again the course of evolution and geological history would have changed under the influences of the lower gravity, the thinner atmosphere, the reduced erosion by falling water, and the probably increased level of background radiation due to more crustal radioactivity and solar cosmic particles. Would evolution have proceeded more rapidly? Would the progression from sea to land and the entry of animal forms into the ecological niches open to airborne species have occurred earlier? Undoubtedly animal skeletons would be lighter, and trees would be generally taller and more spindly.

What if the inclination of the Earth's equator initially had been 60° instead of 23.5°? Seasonal weather changes would then be all but intolerable,

and the only climatic region suitable for life as we
know it would be in a narrow belt within about
5° of the equator. With such a narrow habitable
range, it is probable that life would have had difficulty getting started and, once started, would
have evolved but slowly.

Starting out with an equatorial inclination of
0° would have influenced the course of development of the Earth's life forms in only a minor
way. Seasons would be an unknown phenomenon;
weather would undoubtedly be far more predictable and constant from day to day. The temperate
zones would enjoy a constant spring. However,
the region within 12° of the equator would become too hot for habitability (at least near sea
level) though, in partial compensation, some regions closer to the poles would become more
habitable than they are now. But what about men-
tal and emotional development? In the absence of
seasons, would the environment be less stimulat-
ing, and would a diminution of environmental
pressures have led to less rapid evolution of the
human brain? And if that were not so, how would
the lack of the existence of alternations of a growing season and a non-growing season, a cycle of life
and death and rebirth, have altered man's religious
development?

Suppose the Earth's mean distance from the
Sun were 10 per cent less than it is at present.
Less than 20 per cent of the surface area would
then be habitable. There would be two narrow

regions favorable to life between latitudes 45°
and 65° N and S, separated by a wide and intolerably hot barrier. Land life could evolve independently in these two regions. The polar ice
would not be present, so the ocean level would be
higher than it is now, thus decreasing the land
area.

If the Earth were 10 per cent farther away
from the Sun, the ice-caps would grow, lowering
the sea level. The habitable regions would be
those within 47° of the equator, so that Canada,
Great Britain, Scandinavia, and the Soviet Union
would be frozen out.

If the Earth's rotation rate were increased so
as to make the day 3 hours long instead of the
usual 24, the oblateness would be pronounced,
and changes of gravity with changes of latitude
would be a common part of a traveler's experience.
Day-to-night temperature differences would become small, and it is difficult to predict what
sleeping habits we might have developed or failed
to develop.

On the other hand, if the Earth's rotation rate
were slowed down to make the day 100 hours in
length, day-to-night temperature changes would
be extreme, and weather cycles would have a more
pronounced day-fitting pattern. The Sun's move-
ment across the sky would be almost impercepti-
ble, and few life forms on land could tolerate
either the heat of the long day or the cold of the
long night.

The effects of reducing the eccentricity of the Earth's orbit to 0 from its present value of 0.0167, or increasing it to 0.2, without altering the length of the major axis of the ellipse, would have little effect on the planet.

Increasing the mass of the Sun by 20 per cent, and moving the Earth's orbit out to 1.408 A.U. (130 million miles) to keep the illuminance at the present level, would increase the period of revolution to 1.54 years (563 days) and decrease the Sun's apparent angular diameter from the 32 minutes of arc it is now to 26 minutes of arc. (Since a more massive star is a hotter star, the Sun would not have to exhibit as much apparent surface to give us our necessary illuminance.) The Sun would, under these conditions, be a class F5 star with a total main sequence lifetime of about 5.4 billion years. If the age of the Solar system is taken as 4.5 billion years, then the Earth, under these conditions, could look forward to another billion years (possibly less) of history, rather than the 5 to 7 billion to which it can look forward now. An F5 star may well be more active than our Sun, thus producing a higher exosphere temperature in the planetary atmosphere; but this subject is so little understood at present that no conclusions can be drawn. Presumably, apart from the longer year, the smaller apparent size of the Sun, its more pronounced whiteness, and the "imminence" of doom, life could be much the same.

If the mass of the Sun were reduced by 20

per cent (this time decreasing the Earth's orbital dimensions to compensate), the new orbital distance would be 0.654 A.U. (about 60 million miles). The year's length would then become 0.59 years (215 days), and the Sun would look much larger, with an apparent angular diameter of 41 minutes of arc. The primary would be of spectral type G8 (slightly yellower than our Sun is now), with a main sequence lifetime in excess of 20 billion years. The ocean tides due to the Sun would be about equal to those due to the Moon, so that the rotation rate would have slowed down more than it has, and the day would be more than 24 hours long.

What if the Moon had been located much closer to the Earth than it is—say, about 95,000 miles away instead of 239,000 miles? The tidal braking force by now would probably have been sufficient to halt the rotation of the Earth with respect to the Moon, and the Earth's day would equal its month and become 166 hours (6.9 24-hour days) in length with respect to the Sun. Consequently, the Earth would have great extremes of temperature and would be non-habitable. Increasing the mass of the Moon to 10 times the present value would have the same effect, even though it were left at its present distance. The day-month would then be equal to 26 24-hour periods.

Moving the Moon farther away than it is would have much less profound results: The month would merely be longer and the tides lower. Decreasing the Moon's mass at its present distance would affect only the tides.

What if the properties of some of the other planets of the Solar system were changed? Suppose the mass of Jupiter were increased 1050 times, making it essentially a replica of the Sun. The Earth could still occupy its present orbit around the Sun, but its sky would be enriched by the presence of an extremely bright star of magnitude —23.7, which would supply up to 6 per cent as much heat and light as the Sun. Mercury and Venus could also keep their present orbits, though Mars and Saturn could not. Uranus, Neptune, and Pluto might have modified orbits revolving about the center of mass of the two stars, rather than about one star alone.

All in all, the Earth is a wonderful planet to live on, just the way it is. Almost any change in its physical properties, position, or orientation would be for the worse from our human-centered viewpoint. We are not likely to find a planet that suits us better, although at some future time there may be men who prefer to live on other planets.

And at the present time, since the Earth is as yet the only home we have, we would do well to conserve its treasures and to use its resources intelligently.

Space Flight and
Human Destiny

In the next few centuries, man will be living on
the Earth under conditions of increasing discom-
fort. The population of the Earth is growing at a
rate between 1.5 and 2 per cent per year. It cannot continue to do so indefinitely, and an upper
limit must be reached somehow within the next
several hundred years (hopefully by some means
other than a man-made catastrophe). The final
stabilized population will be considerably higher
than today's population, and the Earth will be
much more crowded than it is now. The incentives
for pioneers to seek new lives among the stars for
themselves and their families will increase continually; and, eventually, the number of human
beings living elsewhere than on Earth may exceed

(even far exceed) the population of the home
planet.

But space flight will not have the effect of reducing our population. This is obvious, of course,
from the present rate of increase in the world's
population. In mid-1962, the world's population
was estimated to have reached the 3-billion mark,
and the net annual rate of increase was estimated
at 1.8 per cent, or over 50 million people per year.
Just to hold the Earth's population constant at the
present time would require the emigration of almost 150,000 people per day—clearly not a reasonable concept. In another century, if the present
rate of natural increase continues, the emigration
rate would have to be stepped up to 900,000 per
day to keep the Earth's population constant at 18
billion people.

It is not the reduction of Earth's population
that will make space flight the most significant
development in the history of civilized mankind,
then. Instead, it will be the gradual multiplication
of human beings living on planets other than the
Earth. Over the generations, man may leave Earth
in considerable numbers and penetrate the Galaxy
to considerable depth. If man learns to travel
through space at, say, one-quarter or one-half
the speed of light, then, even with long planetary
stopovers on his star-hopping expeditions—stopovers long enough to reduce his net advance to
"only" one-tenth the speed of light—the entire
Galaxy could be explored and all its habitable

planets settled within the next million years. Unquestionably, many technological advances will occur before so many years have passed, and the spread of mankind throughout the Galaxy may take place even more rapidly.

And the significance? We have already spoken (see page 210) of the hastened physical evolution of men who colonize planets not exactly like the Earth in physical properties. There may, however, be another form of evolution, more subtle in its nature and more profound in its effects, in the very nature of space flight.

Each stage in the progress of man as he star-hops into new unexplored regions of the Galaxy will be accompanied by an important kind of distillation process. Always, those volunteering for the next expedition into the unknown will tend to be adventurous, self-reliant, inquisitive, courageous, and hardy pioneers, while those selected to go will be chosen on the basis of good health, high professional competence, emotional stability, reliability of judgment, and so on. In the main, these characteristics will be passed on to their descendants, so that a kind of selection process will take place, with those at the frontier of the wave through the Galaxy always representing some of the best qualities of mankind, and leavening all of mankind with those qualities.

Space flight, in short, may well represent a new form of evolutionary pressure—both with respect to the new environments to which man will be exposed and to the new requirements made of his mind and character—a pressure more strenuous than any ever known on Earth.

Its results we scarcely dare imagine—and we regret we cannot live to see them.

Glossary

Accretion. The process of growth by the external addition of new matter and the coherence of separate particles.

Albedo. The ratio that the light reflected from an unpolished surface bears to the total light falling upon it.

Apastron. That point in the orbit of a double star system where the two stars are farthest apart; opposed to periastron.

Aphelion. The point of a planet's orbit most distant from the Sun; opposed to perihelion.

Asteroid. One of the numerous small planets most of whose orbits lie between those of Mars and Jupiter.

Astronomical Unit. Average distance from the Earth to the Sun; abbreviated A.U. One astronomical unit equals about 92.9 million miles.

Binary star system. Two stars relatively close together and revolving about their common center of gravity. The stars revolve in elliptical orbits with periods ranging from a few hours to thousands of years.

Eccentricity (as applied to an elliptical orbit). The ratio of the distance between the center and either focus to the semimajor axis. The eccentricity of a circle is zero; the eccentricity of a parabola, the limiting case of an ellipse, is one.

Ecliptic. The plane of the Earth's orbit.

Ecology. That branch of biology that deals with the mutual relations among organisms and between them and their environment.

Ecosphere. As used here, a region in space in the vicinity of a star in which suitable planets can have surface conditions compatible with the origin, evolution to complex forms, and continuous existence of land life, and surface conditions suitable for human beings and the ecological complex on which they depend.

Equinox. The time at which the Sun crosses the celestial equator; then the day and the night are of equal length.

Escape velocity. The speed that an object must acquire to escape from a planet's gravitation.

Exosphere. The outermost layer of a planet's atmosphere from which gases could escape to space if their molecular velocities were sufficiently high.

Flares (specifically, solar flares). Very bright, spotlike outbursts on the Sun, generally observed over or near large sunspots. Flares occur at unpredictable intervals, last from a few minutes to an hour or more, and emit high-energy protons which constitute one of the more serious hazards of manned space flight.

Forbidden region. An annular band more or less centered on a planet's orbit within which other less massive bodies cannot exist on stable planetary orbits because of perturbing effects produced by the more massive body.

Galaxy. A large gravitational system of stars. Hundreds of thousands of galaxies have been photographed. *Our* Galaxy, the Milky Way, of which our Sun is a member, includes all the stars that can be seen by observers on Earth without the aid of a telescope and most of the objects that can be seen through small telescopes. Our Galaxy is a spiral galaxy, having the general shape of a lens. It is about 100,000 light years in diameter and 10,000 light years thick. The Sun is about two-thirds the distance, or 30,000 light years, from its center. Our Galaxy contains on the order of 100 billion stars.

General planetology. A branch of astronomy that deals with the study and interpretation of the physical and chemical properties of planets.

Geocentric. Relating to the Earth as a center or central point of reference.

Globular cluster. A group of stars clustered into a spherical or slightly flattened spheroidal shape, generally containing thousands of individual stars. The globular clusters, of which about 100 are known, are distributed spherically around the center of our Galaxy.

Gravitation. The universal attraction exerted by every particle of matter on every other particle.

Gravity. The net effect, on the surface of a celestial body, of its gravitation and of the centrifugal force produced by its rotation. On the Earth's surface the value of gravity (symbol g) is about 32 feet/sec^2.

Heliocentric. Relating to the Sun as a center or as a central point of reference.

Hypoxia. Oxygen deficiency in the blood, cells, or tissues.

Illuminance (illumination). The density of luminous flux on a surface; time rate of flow of visible light per unit of surface area.

Inclination. The angle between two planes, such as the angle between the plane of a planet's equator and the plane of its orbit or the angle between the plane of a planet's orbit and a reference plane.

Latitude. Geocentric latitude is the angle between the equatorial plane and a line from the center of the Earth passing through the place; geodetic latitude is the angle between the equatorial plane and the local vertical. Geocentric latitude and geodetic latitude are the same on a sphere, but different on an oblate spheroid.

Light year. The distance over which light can travel in a year's time. Used as a unit in expressing stellar distances. One light year = 0.306 parsec = 6.33×10^4 A.U. = 5.88×10^{12} miles.

Luminosity. As applied to stars, the luminosity is the ratio of the amount of light that would reach us from a star to the amount that would reach us from the Sun if both the star and the Sun were at the same distance from us.

Magnitude. The measure of the relative brightness of a star. The absolute magnitude is the magnitude of a star as it would appear if viewed from the standard distance of 10 parsecs (32.6 light years). The apparent magnitude is its brightness as we see it. The absolute magnitude of the Sun is +4.8; its apparent magnitude is −27.

Main sequence stars. The stars that are in the smooth curve called main sequence of the Hertzsprung-Russell diagram of absolute magnitude versus spectral class; these stars are believed to be in the stable phase of their lifetimes. (After they have consumed a certain fraction of their nuclear fuel, they become unstable and go into later evolutionary stages. Their temperatures, diameters, and internal processes change rapidly, and they become red giants and variables. Finally when their nuclear fuels are exhausted, they become white dwarfs, possibly after having explosively ejected part of their mass.)

Mass ratio. In a system containing two massive bodies, the mass ratio is the ratio of the mass of the smaller to the sum of the masses of the two.

Mean free path. The average distance that a particle (for example, molecule) travels between successive collisions with other particles of an ensemble.

Meteorite. A stony or metallic body that has fallen to the Earth from outer space.

Millibar. A unit of pressure used in meteorology, one one-thousandth of a bar; abbreviated mb. A bar is 10^6 dynes/cm^2. Therefore 1 millibar is 10^3 dynes/cm^2. Normal atmospheric pressure at the Earth's surface is 1013 millibars.

Oblateness. State of being flattened or depressed at the poles; the flattening of a spheroid. Numerically it is the difference between the equatorial and polar diameters divided by the equatorial diameter.

Orbit. The path described by a celestial body in its revolution about another under gravitational attraction.

Orbital velocity. The velocity with which a body moves in an orbit.

Parallax. Generally, the apparent difference in the position of a celestial body when viewed from different positions. *Heliocentric parallax* is the angle subtended by the radius of the Earth's orbit as seen from a specified star, usually measured in thousandths of a second of arc.

Partial pressure. The pressure exerted by one component of a gaseous mixture. Partial pressure equals fractional concentration of the component (by volume) times total pressure.

Periastron. The point in the orbit of a double star system where the two stars are closest to each other.

Perihelion. The point of a planet's orbit closest to the Sun.

Period. The time in which a planet or satellite makes a full revolution about its primary.

Photolysis. The breaking up of a chemical compound by the action of radiant energy, especially light.

Photosynthesis. The formation of organic chemical compounds from water and the carbon dioxide of the air in the chlorophyll-containing tissues of plants exposed to light.

Primary. The massive body (as a star) around which another body is orbiting.

Red giant. A member of a class of very large, low density stars that are not on the main sequence.

Revolution. The term generally reserved for orbital motion as opposed to rotation about an axis (for example, the revolution of the Earth about the Sun).

Roche's limit. The distance from the center of a planet, equal to about 2.45 times its radius, within which a liquid satellite of the same density would be broken apart by the tide-raising forces of the planet.

Roentgen. The unit used in radiology to measure the quantity of absorbed radiation.

Root-mean-square (rms) velocity. The square root of the mean of the squares of the speeds of the particles composing a system.

Rotation. The turning of a body about an axis passed through itself.

Semimajor axis. One-half the longest dimension of an ellipse.

Solstice. One of the two moments in a year when the Sun in its apparent motion attains its maximum distance from the celestial equator.

Spectral classes. Classification of stars based mainly on a progressive change in prominence of certain properties such as color, temperature, and presence and intensity of predominance of certain spectral lines. The principal classes, in descending order of temperature and excitation, are O, B, A, F, G, K, and M. Class O stars are blue-white and very hot; the stars in Class B are also blue-white, but are less hot and are sometimes referred to as helium stars (for the dominant lines in their spectra). Class A contains white stars known as hydrogen stars. Class F stars are yellow-white. Class G stars are yellow (our Sun is a member of this class). Class K stars are orange, and Class M stars are red. Each class is divided into ten spectral types, each designated by a number from 0 to 9 appended to the capital letter denoting the class.

Spectroscopic double. A binary star whose components are too close to be resolved visually but are detected by the mutual shift of their spectral lines owing to their varying velocity in the line of sight.

Synodic. Pertaining to conjunction, especially to the period between two successive conjunctions of the same bodies, as of the Moon or a planet with the Sun.

Universe. All of creation; everything that exists; the entire celestial cosmos.

Velocity of light. Approximately 186,000 miles per second.

Visual double. A binary star that can be separated into two individual stars through use of the telescope.

Vulcanism. A general term for the geological processes in which crustal movements are accompanied by the generation of heat and gases, often with the violent ejection of cinders and lava.

White dwarf. A member of a class of small, very dense white-hot stars of low luminosity, believed to be composed of collapsed degenerate matter and to represent the final stages in the process of stellar evolution when all the nuclear fuel has been used up. White dwarfs have been called "dying stars" that are cooling off and shine only by virtue of the heat generated in their final gravitational collapse.

Related Reading
Material

ALLAN, C. W. *Astrophysical Quantities.* London: The Athlone Press, 1955.

BLANCO, V. M., and McCUSKEY, S. W. *Basic Physics of the Solar System.* Reading: Addison-Wesley Publishing Co., Inc., 1961.

BULLARD, EDWARD. "The Interior of the Earth," in *The Earth as a Planet,* ed. G. P. KUIPER. Chicago: University of Chicago Press, 1954, pp. 57-137.

BUSSARD, R. W. "Galactic Matter and Interstellar Flight," *Astronaut. Acta,* **6,** No. 4 (1960), pp. 179-194.

CALVIN, M. *Chemical Evolution and the Origin of Life,* University of California Radiation Laboratory Report No. UCRL-2124 rev., August 11, 1955.

———. *Origin of Life on Earth and Elsewhere,* University of California Radiation Laboratory Report No. UCRL-9005, December, 1959.

CHAMBERLAIN, J. W. "Upper Atmospheres of the Planets," *Astrophys. J.,* **136,** No. 2 (September, 1962), pp. 582-593.

COCCONI, G., and MORRISON, P. "Searching for Interstellar Communications," *Nature,* **184,** No. 4690 (September 19, 1959), pp. 844-846.

DARWIN, C. G. *The Next Million Years.* Garden City: Doubleday & Company, Inc., 1952.

DAVIS, M. H. "Properties of the Martian Atmosphere," *Quarterly Technical Progress Report (4),* The RAND Corporation, RM-2816-JPL, June 30, 1961.

DE MARCUS, W. C. "Planetary Interiors," *Encyclopedia of Physics,* **52,** ed. S. FLÜGGE. Berlin: Springer-Verlag, 1958.

DOLE, S. H. "Limits for Stable Near-circular Planetary or Satellite Orbits in the Restricted Three-body Problem," *ARS J.*, **31**, No. 2 (February, 1961), pp. 214-219.

FIRSOFF, V. A. *Our Neighbour Worlds*. London: Hutchinsons and Co., Ltd., 1954.

HENDERSON, L. J. *The Fitness of the Environment*. Boston: Beacon Press, 1958.

HOROWITZ, N. H. "The Origin of Life," in *Frontiers of Science*, ed. E. HUTCHINGS, JR. New York: Basic Books, Inc., 1958.

HOYLE, F. *Frontiers of Astronomy*. New York: Harper and Brothers, 1955.

————. *The Nature of the Universe*. New York: Harper and Brothers, 1950.

HUANG, S. "Occurrence of Life in the Universe," *Am. Scientist*, **47**, No. 3 (September, 1959), pp. 397-402.

JEANS, J. H. *Astronomy and Cosmogony*. London: Cambridge University Press, 1929.

JONES, H. SPENCER. *Life on Other Worlds*. New York: New American Library of World Literature, Inc., 1949.

KUIPER, G. P. (ed.). *The Earth as a Planet*. Chicago: University of Chicago Press, 1954.

————. "On the Origin of the Solar System," in *Astrophysics*, ed. J. A. HYNEK. New York: McGraw-Hill Book Co., Inc., 1951.

MACDONALD, G. A. "Volcanology," *Science*, **133**, No. 3454 (March 10, 1961), pp. 673-679.

MARKHAM, S. F. *Climate and the Energy of Nations*. New York: Oxford University Press, 1947.

NEWBURN, R. L., JR. "The Exploration of Mercury, the Asteroids, the Major Planets and Their Satellite Systems, and Pluto," in *Advances in Space Science and Technology*, ed. F. I. ORDWAY, Vol. 3. New York: Academic Press, Inc., 1961.

OPARIN, A. I. *The Origin of Life on the Earth* (3d ed.). New York: The Macmillan Company, 1957.

SÄNGER, E. "Die Erreichbarkeit der Fixsterne," *Proceedings of the VII International Astronautical Congress, Rome, 17-22 September 1956*. Rome: Associazione Italiana Razzi, 1956, pp. 97-113. (*The Attainability of the Fixed Stars*, translated by R. Schamberg, The RAND Corporation, T-69, December, 1956.)

SÄNGER, E. "Nuclear Rockets for Space Flight," *Astronaut. Sci. Rev.*, **3**, No. 3 (July-September, 1961), pp. 9-15.

SHAPLEY, H. *Of Stars and Men.* Boston: Beacon Press, 1959.

SPECTOR, W. S. (ed.). *Handbook of Biological Data.* Wright Air Development Center, Wright-Patterson Air Force Base, Dayton, Ohio, October, 1956.

SPENCER, D. F., and JAFFE, L. D. *Feasibility of Interstellar Travel.* California Institute of Technology Jet Propulsion Laboratory Technical Report No. 32-233, March, 1962.

STRUGHOLD, H. "The Ecosphere of the Sun," *Avia. Med.*, **26**, No. 4 (August, 1955), pp. 323-328.

TAX, S. (ed.). *The Evolution of Life.* (*Evolution after Darwin*, Vol. 1.) Chicago: University of Chicago Press, 1960.

UREY, H. C. "The Atmospheres of the Planets," *Encyclopedia of Physics,* **52**, ed. S. FLÜGGE. Berlin: Springer-Verlag, 1959.

————. *The Planets, Their Origin and Development.* New Haven: Yale University Press, 1952.

VAN DE KAMP, P. "Visual Binaries," *Encyclopedia of Physics,* **50**, ed. S. FLÜGGE. Berlin: Springer-Verlag, 1958.

VAN DEN BOS, W. H. "The Visual Binaries," in *Vistas in Astronomy,* Vol. II, ed. ARTHUR BEER. New York: Pergamon Press, Inc., 1956, pp. 1035-1039.

WALD, G. "Life and Light," *Sci. Am.*, **201**, No. 4 (October, 1959), pp. 92-108.

————. "The Origin of Life," *Sci. Am.*, **191**, No. 2 (August, 1954), pp. 44-53.

Index

About the Authors

STEPHEN H. DOLE is a member of the American Institute of Aeronautics and Astronautics, the American Astronautical Society and the Astronomical Society of the Pacific, and is secretary of the interagency Working Group on Extraterrestrial Resources. He is on the staff of the RAND Corporation, where he has written numerous reports and papers dealing with planetary environments, space photometry, bioastronautics, inorganic methods of oxygen recovery in life-support systems, the space environment, simulated gravity in spacecraft, high energy rocket and aircraft fuels and celestial mechanics.

ISAAC ASIMOV was born in the U.S.S.R. in 1920 and came to the United States at the age of three. He received his B.A., M.A. and Ph.D. from Columbia University. Since 1949 Mr. Asimov has taught at the Boston University School of Medicine, where he is now Associate Professor of Biochemistry. He has written more than fifty books, mostly on scientific subjects. His titles include, *Life and Energy*, *Intelligent Man's Guide to Science*, *Breakthroughs in Science* and *Kingdom of the Sun*.